AF876132

Graphitic Carbon Nitrides

Graphitic carbon nitrides (g-C_3N_4) have attracted much attention as visible-light active photocatalysts. This book focuses on the synthesis, structure modifications, and applications of these advanced carbon materials.

Key Features:

- Summarizes the basic structure, texture, and similarities with carbon materials.
- Explores applications in water splitting, H_2O_2-production, CO_2-conversion to low-molecular-weight hydrocarbons, and degradation and removal of environmental pollutants.
- Examines visible-light irradiation and emphasizes structure modifications of pristine g-C_3N_4, including doping by single heteroatoms and/or molecules.
- Covers the formation of g-C_3N_4-based composites with heavy metals and co-catalysts.
- Describes the conversion of g-C_3N_4 to diamond under high pressure.

Authored by leading authorities, this reference work offers the most thorough review of graphitic carbon nitrides available to aid materials engineers in harnessing these materials for a myriad of photocatalytic applications.

Michio Inagaki is Professor Emeritus at Hokkaido University, Japan. He served as President of the Carbon Society of Japan and was formerly Professor in the Faculty of Engineering at Aichi Institute of Technology, Hokkaido University, Toyohashi University of Technology, Japan, and Visiting Professor of Tsinghua University, China. He received his PhD from Nagoya University.

Masahiro Toyoda is Professor Emeritus at Oita University, Japan, where he had served as Dean of the Faculty of Science and Technology and Director of the Graduate School of Engineering. He received his PhD from Hokkaido University.

Yasushi Soneda is Principal Research Manager at the Energy Process Research Institute, Department of Energy and Environment, National Institute of Advanced Industrial Science and Technology, Japan. He serves as President of the Carbon Society of Japan. He received his PhD from Hokkaido University.

Emerging Materials and Technologies

Series Editor: Boris I. Kharissov

The *Emerging Materials and Technologies* series is devoted to highlighting publications centered on emerging advanced materials and novel technologies. Attention is paid to those newly discovered or applied materials with potential to solve pressing societal problems and improve quality of life, corresponding to environmental protection, medicine, communications, energy, transportation, advanced manufacturing, and related areas.

The series takes into account that, under present strong demands for energy, material, and cost savings, as well as heavy contamination problems and worldwide pandemic conditions, the area of emerging materials and related scalable technologies is a highly interdisciplinary field, with the need for researchers, professionals, and academics across the spectrum of engineering and technological disciplines. The main objective of this book series is to attract more attention to these materials and technologies and invite conversation among the international R&D community.

Handbook of High Entropy Alloys: Fundamentals to Applications
Edited by Shashanka Rajendrachari

Emerging Materials for Biofuel Developments
Edited by Selvakumar Periyasamy, Eduardo Alberto López-Maldonado, and R. Balachandran

Nanostructured Materials for Energy Applications
Edited by Abdullah M. Al-Enizi, Mohd Ubaidullah, Mohd Shkir, and Abhay Kumar Singh

Boron Nitride Nanostructures: Applications in Energy and Healthcare
Indranil Lahiri, Amrita De Adhikari, Souvik Ghosh, Rita Joshi, Satish Jaiswal, and Debrupa Lahiri

Smart Polymers: An Emerging Vista for Nanotheranostics
Dipankar Chattopadhyay, Adrija Ghosh, Jonathan Tersur Orasugh, and Suprakas Sinha Ray

Lightweight Materials for Electric Vehicles: Sustainable Materials, Production Process, and Modeling Techniques
Edited by R. Suresh, C. Durga Prasad, Satish Kumar, Bharath K.N., and Ajith G. Joshi

Graphitic Carbon Nitrides
Michio Inagaki, Masahiro Toyoda, and Yasushi Soneda

For more information about this series, please visit www.routledge.com/Emerging-Materials-and-Technologies/book-series/CRCEMT

Graphitic Carbon Nitrides

Michio Inagaki, Masahiro Toyoda,
and Yasushi Soneda

CRC Press
Taylor & Francis Group
Boca Raton London New York

CRC Press is an imprint of the
Taylor & Francis Group, an **informa** business

Designed cover image: Michio Inagaki, Masahiro Toyoda and Yasushi Soneda

First edition published 2026
by CRC Press
2385 NW Executive Center Drive, Suite 320, Boca Raton FL 33431

and by CRC Press
4 Park Square, Milton Park, Abingdon, Oxon, OX14 4RN

CRC Press is an imprint of Taylor & Francis Group, LLC

© 2026 Michio Inagaki, Masahiro Toyoda and Yasushi Soneda

Reasonable efforts have been made to publish reliable data and information, but the author and publisher cannot assume responsibility for the validity of all materials or the consequences of their use. The authors and publishers have attempted to trace the copyright holders of all material reproduced in this publication and apologize to copyright holders if permission to publish in this form has not been obtained. If any copyright material has not been acknowledged please write and let us know so we may rectify in any future reprint.

Except as permitted under U.S. Copyright Law, no part of this book may be reprinted, reproduced, transmitted, or utilized in any form by any electronic, mechanical, or other means, now known or hereafter invented, including photocopying, microfilming, and recording, or in any information storage or retrieval system, without written permission from the publishers.

For permission to photocopy or use material electronically from this work, access www.copyright.com or contact the Copyright Clearance Center, Inc. (CCC), 222 Rosewood Drive, Danvers, MA 01923, 978-750-8400. For works that are not available on CCC please contact mpkbookspermissions@tandf.co.uk

For Product Safety Concerns and Information please contact our EU representative GPSR@taylorandfrancis.com. Taylor & Francis Verlag GmbH, Kaufingerstraße 24, 80331 München, Germany.

Trademark notice: Product or corporate names may be trademarks or registered trademarks and are used only for identification and explanation without intent to infringe.

ISBN: 978-1-032-96322-8 (hbk)
ISBN: 978-1-032-96323-5 (pbk)
ISBN: 978-1-003-58905-1 (ebk)

DOI: 10.1201/9781003589051

Typeset in Times
by codeMantra

Contents

Preface

The present authors (Michio Inagaki, Masahiro Toyoda, and Yasushi Soneda) have been working mainly on carbon materials, natural and artificial graphite, carbon fibers, carbon blacks, graphite intercalation compounds, etc., and have published numerous research papers and reviews focusing on a specific problem. The importance of nanotextures of carbon materials has been pointed out, and the nanotexture due to the preferential orientation of strongly anisotropic hexagonal carbon layers (i.e., graphene layers) is first classified into four categories: planar, axial, point, and random orientations. These orientations are fundamental to the carbon materials, such as graphite and various artificial graphites, various carbon fibers, various carbon blacks, various porous carbons, amorphous carbons, and glass-like carbons.

Graphitic carbon nitrides with a stoichiometry of C_3N_4 (abbreviated as *g*-CN and CN) have attracted the attention of scientists and engineers mainly due to their photocatalytic activity in visible light. It has a layered structure similar to that of graphite. The controlling and modifying of the structure and texture of *g*-CN-related materials are known to be the important factors in enhancing their photocatalytic activity. These processes are very similar to those applied on carbon materials, as discussed in detail by the present authors (Michio Inagaki and Masahiro Toyoda) in 2019, and received relatively high citations (approximately 400 according to Google Scholar). In this book, therefore, *g*-CN materials are discussed in terms of their structure, synthesis, texture control, modification, and applications as important factors in understanding the materials science and engineering of the *g*-CN-related materials.

This book consists of nine chapters, focusing on the fundamental structure of *g*-CN, followed by the syntheses of *g*-CN, and then on their applications, i.e., H_2-evolution via water splitting, CO_2-conversion to CO and hydrocarbons, N_2-fixation to NH_3, H_2O_2-production, and pollutant degradation at room temperature under solar light irradiation. For each application, the experimental results reported are summarized, including the recently published results, aiming to assist the scientists and engineers who are planning to work on further advancing investigations, including a scale-up of the *g*-CN production.

Acknowledgments

The authors (Michio Inagaki, Masahiro Toyoda, and Yasushi Soneda) would like to express their sincere thanks to the people who kindly provided the data and figures for this book. The authors also thank all the members who took care of this book at Taylor & Francis.

1 Introduction

1.1 CARBON NITRIDE

Graphitic carbon nitride (g-C_3N_4) possesses a layered structure similar to that of graphite and presents interesting and unique physicochemical properties due to s-triazine cores. However, its electronic and electrochemical applications are limited by its relatively poor conductivity. Its photocatalytic activity for visible light attracts the attention of scientists for hydrogen evolution by water splitting and for pollutant degradation at room temperature under visible-light irradiation. g-C_3N_4 itself and g-C_3N_4-based materials have emerged as promising candidates for a variety of applications, including CO_2-conversion to CO and low-molecular-weight hydrocarbons, H_2O_2-generation, N_2-fixation (N_2-reduction to NH_3), and a substrate for various catalysts. In addition, some of C_3N_4 polymorphs are superhard, comparable, and even superior to diamond.

This book focuses on graphitic C_3N_4 (abbreviated as g-CN) and discusses it in comparison with carbon materials [1]. It has been reported that other crystalline structures of C_3N_4 exist, including α-C_3N_4, β-C_3N_4, etc. [2]. However, they are only mentioned in comparison with g-C_3N_4 in this book. Numerous papers have been published on carbon nitrides, but most of them deal with carbon nitrides, whose elemental compositions are CN_x (where x is less than 1.0) and their basic structures are carbon, in which nitrogen atoms are either substituted for carbon atoms or attached as nitrogen-containing functional groups. CN_x materials with $x > 1.0$ are often referred to in the literature as nitrogen-rich carbon nitrides. Therefore, the materials having the composition of CN_x with $x < 1.0$ are referred to only when they need to be compared with g-CN with $x > 1.0$. The CN_x ($x < 1.0$) materials were previously reviewed as nitrogen-doped carbon materials [3,4]. In this book, g-CN materials are discussed in terms of their structure, synthesis, texture control, modification, and applications as the important factors for understanding the g-CN-related materials, similar to the carbon materials, as explained in detail in a series of books [5–7].

The g-CN was critically reviewed on its starting idea, having similarity with silicon nitride Si_3N_4 [8]. A detailed review was presented on the design and construction of the pristine g-CN and its nanocomposites, including (1) nanotexture control, such as by hard and soft templating, supramolecular re-organization, exfoliation, (2) its functionalization at an atomic level (elemental doping) and molecular level (copolymerization), and (3) its modification with another semiconductor or a metal as a co-catalyst to form heterojunctions to be applied for a variety of reactions contributing to environmental de-contamination [9]. g-CN was also reviewed, how electronic band positions and bandgap can be varied by doping and copolymerization, and how the solid g-C_3N_4 can be textured to make it an effective photocatalyst with heterojunctions [10]. The application of g-CN for CO_2-conversion was focused [11]. It was shown to have a significant potential for photocatalytic pollutant degradation [12].

DOI: 10.1201/9781003589051-1

The visible-light-sensitive water splitting performance of photocatalytic materials was reviewed with special attention to the heterojunctions in the catalysts [13]. The photoelectrochemical application of *g*-CN was also reviewed [14,15]. The effects of compositing and morphology modulation on photocatalytic performances were reviewed [16]. The four most frequently used modification strategies are discussed, including vacancy introduction, doping with nonmetal elements, modifying with nonmetal quantum dots, and coupling with other semiconductors [17]. *g*-C_3N_4 was reviewed as one of the materials for effective utilization of solar energy [18].

1.2 STRUCTURE OF *g*-C_3N_4

The structure of *g*-CN is often considered to be evolved from the compounds with direct bonding of C–N, urea and ethylenediamine, through cyanamide, melamine, and their polymerized derivatives, resulting in either triazine-based or heptazine-based structures, as shown in Figure 1.1. Its basic structural unit (BSU) is either triazine or heptazine cores, both of which are planar, allowing the polymerized final products to adopt a layered structure. This is the main reason why these polymerized products are referred to as "graphitic" carbon nitrides. BSU, whether triazine or heptazine, is planar, while it is not necessarily bonded with neighboring BSUs in the same plane; in other words, neighboring BSUs can be easily rotated, resulting in the difficulty of having long-range order in structure, both in plane and perpendicular to the plane, resulting in amorphous *g*-CN. The intrinsic cavities surrounded by N-atoms (either 3 or 6, V_{3N} and V_{6N}), as highlighted by full circles in Figure 1.1, are characteristic of *g*-CN, which can capture (anchor) the foreign atoms and/or molecules, similar to porphyrin compounds. Furthermore, different vacancies of N and C

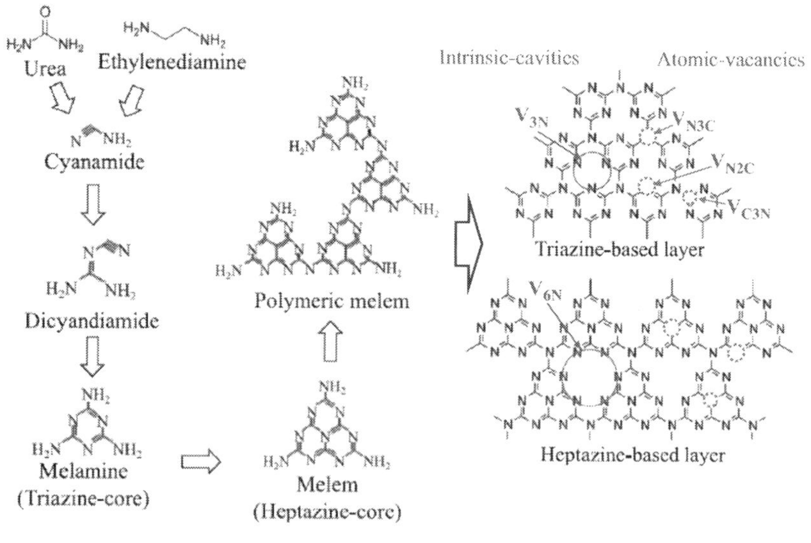

FIGURE 1.1 Structural evolution from urea and ethylenediamine to *g*-CN. Intrinsic cavities surrounded by N-atoms (V_{3N} and V_{6N}) and atomic vacancies (V_{N3C}, V_{N2C}, and V_{C3N}) are highlighted by full and dotted circles, respectively.

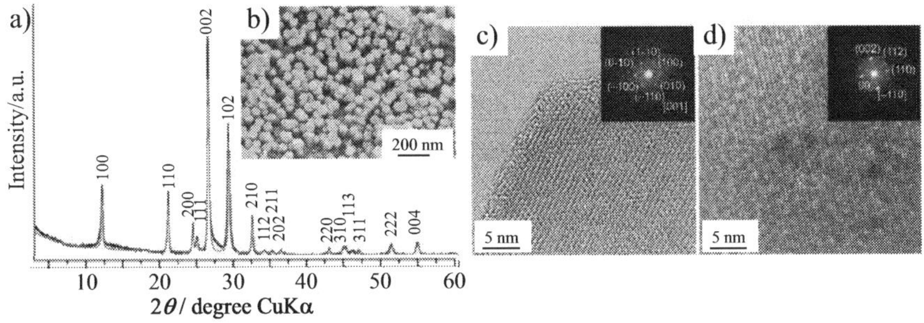

FIGURE 1.2 Crystalline *g*-CN synthesized using eutectic molten salt LiCl/KCl [19].

can occur, N-vacancies coordinated by 3 C-atoms (V_{N3C}), and 2 C-atoms (V_{N2C}), and C-vacancies coordinated by 3 N-atoms (V_{C3N}), as illustrated in Figure 1.1 on triazine- and heptazine-based molecules. These intrinsic cavities and defective vacancies on *g*-CN offer distinct advantages in their applications as the catalysts. These features facilitate the loading of metals as single atoms, working as the photocatalytic active centers, and expanding the possibilities for doping and incorporating a diverse array of foreign atoms and molecules. Introduction of N-vacancies (V_{N3C} and V_{N2C}) into *g*-C_3N_4 has been demonstrated to create defect energy levels and trap electrons, consequently accelerating the separation of e^-/h^+ pairs and effectively boosting photocatalytic activity. Nitrogen vacancies can also serve as adsorption active sites, thereby enhancing the adsorption capacity of the catalyst toward the target molecules.

XRD powder pattern of a *g*-CN synthesized from dicyandiamide ($C_2H_4N_4$, DCDA) by using a eutectic mixture of LiCl/KCl as a solvent is shown in Figure 1.2a–d, together with SEM image and lattice fringe images of transmission electron microscopy (TEM) along [001] and [110] directions [19]. The presence of sharp XRD 00*l*, *hk*0, and *hkl* diffraction peaks, in conjunction with large-range lattice fringes in the two directions, suggests a highly crystalline structure of the *g*-CN. From X-ray diffraction and TEM analyses, the crystal belongs to the hexagonal system with the space group $P6_3cm$ with the lattice parameters of $a_0 = 0.8434$ and $c_0 = 0.6722$ nm. The importance of the crystallinity of *g*-CN for its photocatalytic performances was recently pointed out by reviewing various published literature [20].

In contrast, most of the *g*-CNs prepared from urea, melamine, etc., are much poorer crystalline, of which XRD patterns are composed from only two diffraction peaks, 002 and 10, as shown in Figure 1.3a, which was synthesized from urea at a temperature of 450°C–600°C [21]. With increasing synthesis temperature, the profiles of 002 and 10 diffraction peaks become sharp, suggesting a gradual growth of the BSU (crystallite), i.e., an increase in its thickness, in other words, an increase in the number of stacking CN layers, and an increase in its lateral size of the layers. However, the stacking of CN layers is still not regular because of the characteristically unsymmetric profile of the two-dimensional 10 diffraction peak (so-called turbostratic stacking). Consequently, most of the *g*-CNs synthesized are amorphous, often being called polymeric *g*-CNs. Even after the synthesis at a high temperature above 550°C, only traces of other diffraction lines are recognizable. In Figure 1.3b, XRD patterns are compared on two *g*-CNs, tubular and bulky *g*-CN, both being

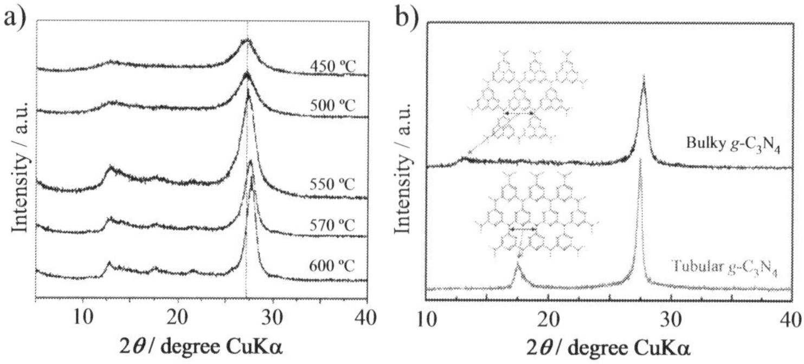

FIGURE 1.3 XRD powder patterns of *g*-CNs: (a) *g*-CNs derived from urea [21] and (b) tubular and bulky *g*-CNs derived from melamine [22].

synthesized from melamine [22]. The former was obtained by sintering the white precipitates from melamine-saturated ethylene glycol solution by adding nitric acid at 350°C for 1 h in air, and the latter was synthesized by directly heating melamine at 520°C for 2 h. The latter (bulky CN) provided a broad 10 line at 2θ of around 13° (interlayer spacing d_{10} of 0.681 nm), as observed in Figure 1.3a, which was reasonably supposed to be due to the distance of the tri-*s*-triazine-based structure, as illustrated in Figure 1.3b as an inset. In contrast, the former (tubular CN) exhibits a 10 diffraction peak at around 17.4° (d_{10} of 0.49 nm), which was reasonably supposed to be due to a *s*-triazine-based structure, as illustrated in Figure 1.3b. It should be noted that the CN nanotubes possess an atom ratio of *C*/*N* of 3/4.82. However, both CNs have a 002 diffraction peak located at around 27.4° (d_{002} of 0.326 nm, proving a graphite-like layer structure), while their BSUs were small.

1.3 SIMILARITIES WITH CARBON MATERIALS

Carbon materials include a wide range of structures composed of hexagonal rings of carbon atoms with sp^2 hybrid bonds associated with π electrons, from very small lateral sizes with few stacked layers to large lateral sizes with many stacked layers, the former being found in amorphous carbons and the latter being represented by graphite [5,6]. Furthermore, diamond is composed of sp^3 hybrid bonds, while carbon nanotubes and fullerenes primarily consist of hexagons with sp^2 bonds, but include pentagons. The structures of amorphous carbon to graphite are controlled by high temperature treatment up to 3,000°C in inert atmosphere and are often characterized by lattice parameters, such as interlayer spacing d_{002} and crystallite sizes of L_c and L_a, which are determined by the analyses of XRD patterns. Graphite can be transformed into diamond directly under high temperature and high pressure (Table 1.1).

1.4 CONSTRUCTION OF THIS BOOK

The present book, *Graphitic Carbon Nitrides g-C_3N_4*, consists of nine chapters, focusing on the fundamental structure of graphitic carbon nitride (g-C_3N_4) (**Chapter 1**), followed by the synthesis of g-C_3N_4 in **Chapter 2**, and on their applications in

TABLE 1.1
Comparison to Carbon Materials in Synthesis, Structure, Morphology, and Properties

Characteristics		Carbon Nitrides	Carbon Materials
Atomic composition		C_3N_4	C
Basic structure		Carbon and nitrogen hexagonal rings Triazine and/or heptazine cores	Carbon hexagonal rings
Syntheses		Mild temperature (500°C–600°C) in air	High temperatures (>1,000°C) in inert atmosphere
Crystalline structure	Under atmospheric pressure	Graphitic layered structure Hexagonal and rhombohedral	Graphite (layer structure), fullerene, and carbyne Hexagonal and rhombohedral
	Under high pressure	Transform to rhombohedral structure. Decomposition to diamond (carbon) and N_2 gas	Transform to diamond (cubic)
Particle morphology		Lamellar, tubular, and foam	Lamellar, fibrous, tubular, and foam
Texture control	Templating	Hard templates (mesoporous silica, colloidal silica, etc.)	Hard template (zeolite, SiO_2, MgO, etc.)
	Exfoliation	Thermal exfoliation	Thermal, microwave irradiation, electrochemical, surfactant-assisted exfoliation
	Doping	B, extra C, P, S, K, LiCl, etc.	B, N, etc.
Composite formation	Self-composite	Coupling of *g*-CN with different crystallinities and with doped *g*-CNs	Carbon–carbon composites, such as carbon fiber with carbon
	With various species	Compositing with various species (metals, inorganics, and organics)	Compositing with various species (metals, inorganics, and organics)
Properties	Electric	Poor electric conduction	Highly conductive along the layer and less conductive across the layer
	Photo	Visible light (UV-vis light, solar light) active	High conversion of solar energy to thermal energy
	Thermal	Poor thermal conduction	Highly conductive along the layer and less conductive across the layer

Chapters 3–7, H_2-evolution (water splitting), CO_2-conversion (CO_2-reduction), H_2O_2-production, N_2-fixation (NH_3-generation), and pollutant degradation, respectively. In **Chapter 8**, the application of *g*-C_3N_4 as a precursor for the synthesis of

diamond and a polymorph of *g*-C_3N_4 under high temperature and pressure was explained by presenting a phase diagram and the advantages of *g*-C_3N_4 as a precursor. Since C_3N_4 polymorphs have also attracted attention as members of superhard materials, i.e., β-C_3N_4, in addition, the hardness of several C_3N_4 polymorphs was additionally compared with diamond. In **Chapter 9**, the prospects of the photocatalytic activity of *g*-C_3N_4 are discussed. In **Chapters 3–8**, the applications of the *g*-C_3N_4-related materials are described by classifying into two groups: *g*-C_3N_4 and modified *g*-C_3N_4, and *g*-C_3N_4-based composites.

REFERENCES

1. Inagaki M, Tsumura T, Kinumoto T, et al., Graphitic carbon nitrides (*g*-C_3N_4) with comparative discussion to carbon materials. *Carbon* 2019; 141: 580–607.
2. Kroke E, Schwarz M, Novel group 14 nitrides. *Coord Chem Rev* 2004; 248: 493–532.
3. Wu J, Pan Z, Zhang Y, et al., The recent progress of nitrogen-doped carbon nanomaterials for electrochemical batteries. *J Mater Chem A* 2018; 6: 12932–44.
4. Inagaki M, Toyoda M, Soneda Y, et al., Nitrogen-doped carbon materials. *Carbon* 2018; 132: 104–40.
5. Inagaki M, Kang F, *Materials Science and Engineering of Carbon: Fundamentals*. Tsinghua University Press and Elsevier, Beijing, China and Amsterdam, Netherlands, 2014.
6. Inagaki M, Kang F, Toyoda M, et al., *Advanced Materials Science and Engineering of Carbon*. Tsinghua University Press and Elsevier, Beijing, China and Amsterdam, Netherlands, 2013.
7. Inagaki M, Kang F (Eds.), *Materials Science and Engineering of Carbon: Characterization*. Tsinghua University Press and Elsevier, Beijing, China and Amsterdam, Netherlands, 2017.
8. Kroke E, *gt*-C_3N_4 The first stable binary carbon(IV) nitride. *Angew Chem Int Ed* 2014; 53: 11134–6.
9. Ong W-J, Tan L-L, Ng YH, et al., Graphitic carbon nitride (*g*-C_3N_4)-based photocatalysts for artificial photosynthesis and environmental remediation: Are we a step closer to achieving sustainability? *Chem Rev* 2016; 116: 7159–329.
10. Wang Y, Wang XC, Antonietti M, Polymeric graphitic carbon nitride as a heterogeneous organocatalyst: From photochemistry to multipurpose catalysis to sustainable chemistry. *Angew Chem Int Ed* 2012; 51: 68–89.
11. Lu Q, Eid K, Li W, et al., Engineering graphitic carbon nitride (*g*-C_3N_4) for catalytic reduction of CO_2 to fuels and chemicals: Strategy and mechanism. *Green Chem* 2021; 23: 5394–428.
12. You JK, Xiao M, Liu SQ, et al., How carbon contamination on the photocatalysts interferes with the performance analysis of CO_2 reduction. *J Mater Chem A* 2023; 11: 10149–54.
13. Afroz K, Moniruddin M, Bakranov N, et al., A heterojunction strategy to improve the visible light sensitive water splitting performance of photocatalytic materials. *J Mater Chem A* 2018; 6: 21696.
14. Zou X, Sun Z, Hu YH, *g*-C_3N_4-based photoelectrodes for photoelectrochemical water splitting: A review. *J Mater Chem A* 2020; 8: 21474–502.
15. Alzakia FI, Tan SC, Liquid-exfoliated 2D materials for optoelectronic applications. *Adv Sci* 2021; 8: 2003864.
16. He F, Wang Z, Lia Y, et al., The nonmetal modulation of composition and morphology of *g*-C_3N_4-based photocatalysts. *Appl Catal B* 2020; 269:11828.
17. Tang C, Cheng M, Lai C, et al., Recent progress in the applications of non-metal modified graphitic carbon nitride in photocatalysis. *Coord Chem Rev* 2023; 474: 214846.

18. Soneda Y, Toyoda M, Inagaki M, Contribution of carbon materials to effective utilization of solar energy: A review. *Sus Mater Tech* 2025; 45: e01537.
19. Bojdys MJ, Muller J, Antonietti M, et al., Ionothermal synthesis of crystalline, condensed, graphitic carbon nitride. *Chem Eur J* 2008; 14: 8177–82.
20. Li H, Cheng B, Xu J, et al., Crystalline carbon nitrides for photocatalysis. *EES Catal* 2024; 2: 411–47.
21. Dong F, Wu LW, Sun YJ, et al, Efficient synthesis of polymeric g-C_3N_4 layered materials as novel efficient visible light driven photocatalysts. *J Mater Chem* 2011; 21:15171–4.
22. Gao J, Zhou Y, Li Z, Y, et al., High-yield synthesis of millimeter-long, semiconducting carbon nitride nanotubes with intense photoluminescence emission and reproducible photoconductivity. *Nanoscale* 2012; 4: 3687–92.

2 Synthesis and Modification of *g*-C_3N_4

g-C_3N_4-based materials are attracting attention mainly because of their photocatalytic activity under visible light (solar light), which are obtained via a low-cost and environment-friendly synthesis process. They are promising for H_2-evolution, H_2O_2-production, CO_2-conversion, pollutant degradation, and N_2-fixation processes under visible-light irradiation. *g*-C_3N_4-based materials must have appropriate composition, structure, and texture to fulfill the requirements of these applications. Numerous processes and treatments have been proposed and investigated.

2.1 SYNTHESES OF *g*-C_3N_4

g-C_3N_4 (*g*-CN) has been synthesized from different precursors by thermal polymerization under relatively mild conditions, such as at low temperatures of 500–650°C in air using a ceramic crucible with a lid, as summarized for some representative precursors [1] in Figure 2.1.

Amorphous *g*-CN (*a*-CN) was prepared by polycondensation of dicyandiamide (DCDA) at 500°C for 4h in static air [2]. Annealing the *a*-CN at 620°C for 2h in an Ar flow yielded the crystalline *g*-CN (*c*-CN). The XRD pattern and UV-vis absorption spectra of the resultant *a*-CN and *c*-CN are compared in Figure 2.2a and b, respectively, revealing that *c*-CN has much higher crystallinity and a wider light absorption range. In Figure 2.2c and d, the proposed structural models of *c*-CN and *a*-CN are illustrated, respectively. The bandgap was calculated to be 1.90 and 2.82 eV

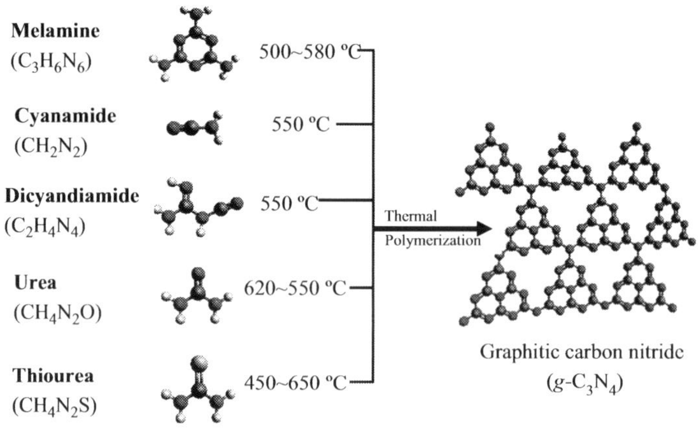

FIGURE 2.1 Schematic illustration of the synthesis process of *g*-C_3N_4 by thermal polymerization of different precursors [1].

DOI: 10.1201/9781003589051-2

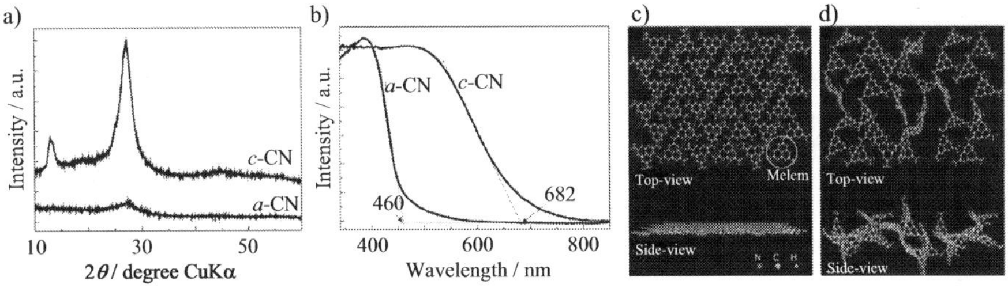

FIGURE 2.2 Amorphous and crystalline g-CNs (a-CN and c-CN, respectively): (a) XRD patterns, (b) UV-vis absorption spectra and the structure models for (c) c-CN and (d) a-CN [2].

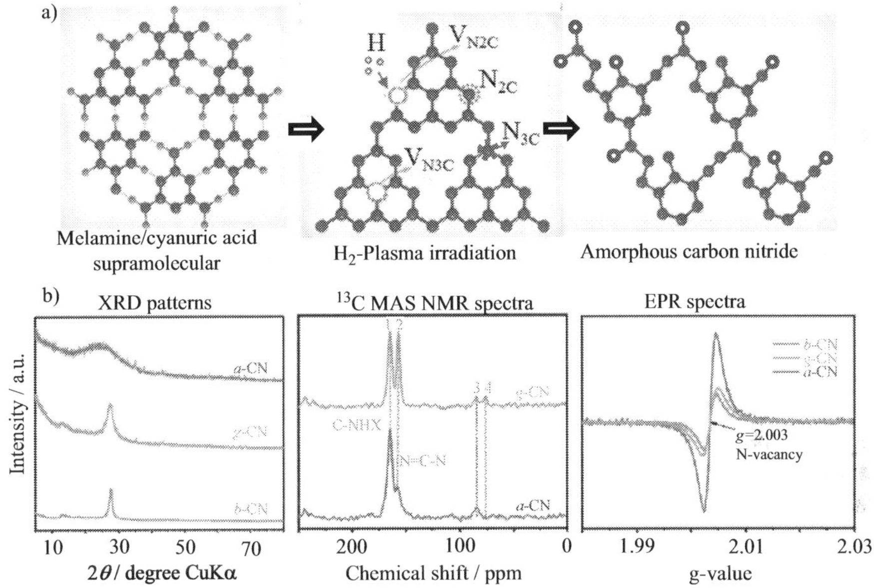

FIGURE 2.3 Amorphous g-CN (a-CN): (a) schematic illustration of the preparation of the a-CN and (b) XRD patterns, ^{13}C solid-state MAS NMR spectra and EPR spectra of a-CN, b-CN, and g-CN [3].

for a-CN and c-CN, respectively. Consequently, the c-CN exhibited much higher photocatalytic H_2 evolution than the a-CN.

Amorphous CN (a-CN) was prepared from melamine/cyanuric acid (CA) supramolecular aggregates using H_2-plasma irradiation [3]. Figure 2.3a illustrates the structural change from supramolecular aggregate to a-CN through H_2-plasma irradiation. During plasma irradiation, N-vacancies and H atoms are introduced, causing some disturbances in the structure. In Table 2.1, elemental composition and pore structure parameters of the a-CN are compared with polymeric graphitic CN (g-CN), which was synthesized by calcining the supramolecular aggregate at 550°C and with bulk CN (b-CN), which was synthesized by calcining melamine at 550°C.

TABLE 2.1
Elemental Composition by XPS Analyses and Pore Structure Parameters of Photocatalysts [3]

	Elemental Composition (at%)					Pore Parameters	
Photocatalyst	**C**	**N**	**O**	**N/C**	**N_{2C}/N_{3C}**	**S_{BET} (m^2/g)**	**V_{total} (cm^3/g)**
a-CN	48.74	47.61	3.65	0.98	2.4	405.758	0.504
g-CN	47.59	49.21	3.2	1.03	3.4	29.416	0.084
b-CN	43.21	53.03	3.76	1.23	3.7	4.891	0.029

The marked reduction in the N_{2C}/N_{3C} ratio of the *a*-CN suggests the formation of N-vacancies. Elemental analyses showed that the *a*-CN contains a high hydrogen content of 2.83%, which is higher than 2.26% found in *g*-CN. In Figure 2.3b, XRD pattern and ^{13}C solid-state MAS NMR and EPR spectra of *a*-CN are compared with those of the *g*-CN and the *b*-CN. The *a*-CN shows distinctly different XRD patterns in comparison with the *b*-CN and *g*-CN. Obviously, the two characteristic diffraction peaks located at 13.1° and 27.5° corresponding to 10 and 002 planes of CN. The ^{13}C NMR spectrum shows the peak at approximately 164.5 ppm (peak 2 in Figure 2.3b) assigned to the C atoms in N=C–N bonds, which is drastically reduced in *a*-CN. Meanwhile, peak 1 assigned to NH_X remains the same in *a*-CN, indicating the existence of N-vacancies. An EPR spectrum, a single Lorentzian line with a *g*-value of approximately 2.003, was observed on three CNs, suggesting that *a*-CN, *g*-CN, and *b*-CN contain only one type of paramagnetic species. This paramagnetic species in the *a*-CN may be attributed to the unpaired electrons in the localized π-conjugated structure, which can be attributed to the N-vacancy. Table 2.1 compares the elemental composition obtained by XPS analysis and the pore structure parameters. The *a*-CN has a low N/C atomic ratio and a small N_{2C}/N_{3C} ratio, mainly due to the introduction of H together with N-vacancies.

N-deficient *g*-CNs (V_N-CN) with different N-vacancy densities were prepared via a simple $NaBH_4$ reduction with varying treatment times [4]. The pristine *g*-CN was synthesized by direct calcination of urea at 550°C for 2 h. The *g*-CN (0.4 g) was mixed with 0.3 g of $NaBH_4$ by dissolving in 15 mL of water and then calcined at 300°C for different times to produce N-deficient CN (coded as V_N-CN-X, where X corresponds to the time of $NaBH_4$-treatment). In Figure 2.4, ^{13}C NMR spectra of V_N-CN-X and the pristine CN are shown together with the proposed structure with N-vacancies. The V_N-CN-X exhibits significant photodegradation removal of ciprofloxacin (CIP). N-vacancies in the centers of heptazine rings of *g*-CN were successfully created via a photochemical-assisted N-etching [5], which delivered a high CO_2-conversion performance.

g-CN was synthesized by heating the powder mixtures of melamine (*M*) with urea (*U*) in different weight ratios (100/0, 75/25, 50/50, 25/75, and 0/100) in an alumina crucible with a lid at 550°C for 2 h [6]. The crystallinity of the resultant *g*-CNs depends strongly on the mixing ratio *M*/*U*, as shown by XRD powder patterns in Figure 2.5a, revealing a gradual sharpening and intensifying of the 002 diffraction line with increasing *M*/*U* weight ratio, while no marked change in its position

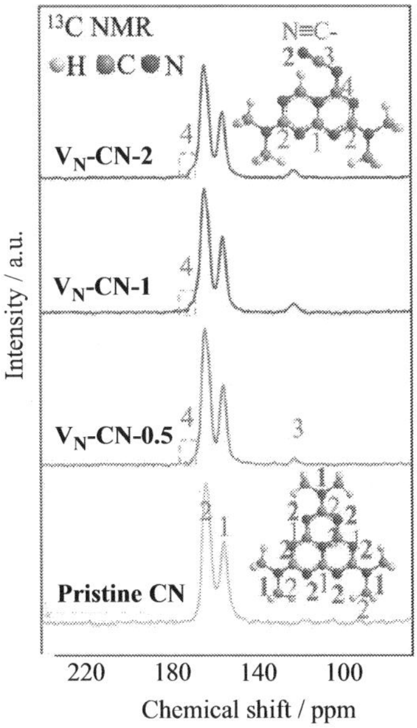

FIGURE 2.4 ^{13}C NMR spectra of the V_N-CN-X and the pristine CN with proposed defective structures (insets) [4].

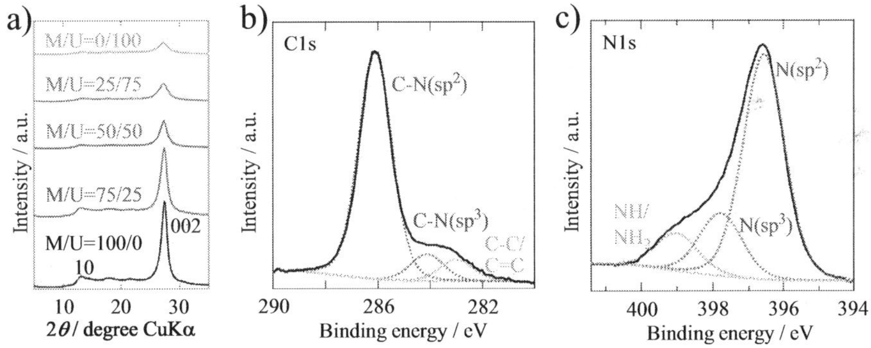

FIGURE 2.5 g-CNs synthesized from the mixtures of melamine (M) and urea (U) in different weight ratios: (a) XRD patterns and XPS spectra for (b) C1s and (c) N1s with de-convolution results on the g-CN obtained from 100/0 (melamine) [6].

(diffraction angle). The change in the 002 peak profile suggests the gradual increase in the thickness of g-CN layers stacking, i.e., an improvement in crystallinity. With increasing M/U ratio, the two-dimensional 10 diffraction peak becomes detectable, but its profile is unsymmetrical, suggesting that the stacking of g-CN layers is disordered, forming a so-called turbostratic structure, as can be seen on low-temperature-treated carbon materials. In Figure 2.5b and c, the XPS spectra for the C1s and N1s are shown, as an example, on the g-CN prepared from a 100/0 mixture (i.e., pure

TABLE 2.2
Elemental Composition and Abundance Ratios of C and N of the *g*-CNs Obtained from Different *M/U* Weight Ratios [6]

M/U	Elemental Composition (mol%)			Abundance Ratio of C			Abundance Ratio of N		
Ratio	C	N	O	C-N(sp^2)	C-N(sp^3)	C-C/C=C	N(sp^2)	N(sp^3)	NH/NH_2
100/0	42.0	55.8	2.2	82.5	9.7	7.8	69.2	19.2	11.6
75/25	41.9	55.9	2.3	84.4	9.2	6.5	72.6	17.5	9.9
50/50	42.9	54.5	2.6	85.3	9.1	5.6	71.9	16.5	11.6
25/75	43.5	54.2	2.4	82.6	11.7	5.7	70.3	17.2	12.4
0/100	42.8	54.7	2.5	78.2	8.6	13.2	66.5	20.4	13.1

melamine). The N1s spectrum can be deconvoluted into N(sp^2) and N(sp^3), where N(sp^2) presents as pyridinic N atoms at the vertices of the heptazine rings and at the center of the rings, and N(sp^3) is a tertiary N atom connecting three carbon atoms. Similarly, the C1s spectrum can be deconvoluted into C–N(sp^2) and C–N(sp^3), as well as single and/or double bonds between carbon atoms (C–C and/or C=C). Table 2.2 summarizes the abundance ratios of each bonding state together with the elemental composition of the *g*-CNs. The abundance ratios of carbon (C) and nitrogen (N) reveal that the chemical structure of *g*-CN prepared from urea (the 0/100 mixture) differs significantly from that of *g*-CN prepared from melamine (100/0 mixture). The photocatalytic degradation performance of formaldehyde in air was studied using the resultant *g*-CNs. The catalyst prepared from the 50/50 mixture exhibited the highest efficiency. *g*-CNs were prepared from the mixture with a *M/U* mixing ratio of 2/98, 10/90, 20/80 and 50/50 by heating at 550°C for 2 h [7]. Structure changes with the mixing ratio *M/U* were discussed on the basis of the measurements using the characterization techniques, including FESEM, AFM, XRD, HRTEM, Raman, FTIR, UV, and PL spectroscopies, BET, and TGA analyses. The stacked micron-sized lamellae of melamine-derived *g*-CN transform into the porous and open structures consisting of small-sized domains in the urea-derived *g*-CN.

Further calcination of polymeric CN (*p*-CN) at a high temperature of 620°C for 2, 4, and 6 h in Ar flow resulted in the creation of N-vacancies in *p*-CN. The resultant CNs were coded as V_N-CN-X, where X represents the re-calcination hours in an Ar atmosphere. These V_N-CN-X could be applied as the photocatalyst for the nitrogen reduction reaction (N_2-fixation) under ambient conditions [8]. The structure of V_N-CN-X is shown by highlighting N-vacancies in Figure 2.6a. The pristine *p*-CN showed relatively sharp 10 and 002 diffraction peaks at 13.08° and 27.38°, respectively, while V_N-CN-X exhibited a very broad 002 diffraction peak but no 10 peak, because of its highly defective structure. In Figure 2.6b, their UV-vis spectra are compared with the pristine *p*-CN, revealing that the introduction of N-vacancies resulted in the creation of new absorbance at the wavelength of around 500 nm, suggesting that the resultant V_N-CN-X is visible-light active.

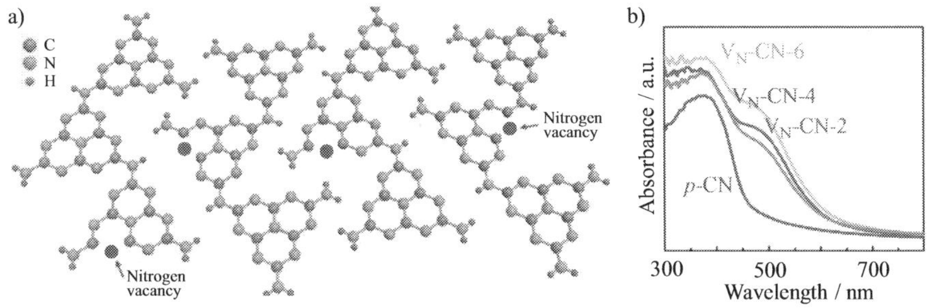

FIGURE 2.6 CN having N-vacancies V_N-CN-*X*: (a) illustration of the structure by highlighting N-vacancies and (b) UV-vis diffuse reflection spectra [8].

TABLE 2.3
Pore Structure of the Modified *g*-CNs Determined by N_2 Adsorption Isotherms [10]

Catalyst	Glycerol/Water (Volume)	S_{BET} (m^2/g)	V_{total} (cm^3/g)	Pore Size (nm)
Gl_0/CN (pristine)	0/60	162.6	0.8	10.4
Gl_{10}/CN	10/50	170.5	0.9	11.6
Gl_{30}/CN	30/30	177.7	1.2	13.3
Gl_{50}/CN	50/10	182.1	1.3	14.7
Gl_{60}/CN	60/0	164.9	1.0	13.9

The supramolecular aggregates of melamine (a *g*-CN precursor) were prepared via hydrothermal treatment (HT) at 140°C of an aqueous solution of melamine (M, 3 g) with different amounts of HCl (0, 4.2, 8.4, or 12.6 mmol). The resultant precursors are coded as M-HCl_0, M-$HCl_{4.2}$, M-$HCl_{8.4}$, and M-$HCl_{12.6}$, respectively [9]. These precursors were converted to homogeneous pastes by mixing with ethylene glycol and then pasted onto a fluorine-doped tin oxide (FTO) substrate and calcined at 530°C in N_2 flow to fully convert the precursors to *g*-CN. The resulting CN-film on a substrate exhibited a 1D configuration, a high specific surface area, and an abundance of defects.

Pore structure of urea-derived *g*-CN powder was controlled using a glycerol/water mixed solvent (0/60, 10/50, 30/30, 50/10, and 60/0 by volume), followed by HT at 160°C for 10 h [10]. The resultant *g*-CNs were categorized according to the amount of glycerol used (X) as Gl_X/CN. In Table 2.3, the pore structure parameters are listed. With increasing *X* (relative amount of glycerol), the pore parameters increase, reaching their maximum at $X=50$ (Gl_{50}/CN). An increase in *X* resulted in a slight red shift in the light absorption capacity in the 420–800 nm region, which was attributed to the introduction of 3C-coordinated nitrogen-vacancies (V_{N3C}). The resulting Gl_X-CNs demonstrated an enhanced CO_2-conversion (CO-evolution) performance.

Annealing the urea-derived *g*-CN in NH_3 at 500°C for 3 h introduced 3N-coordinated carbon-vacancies (V_{C3N}) into the *g*-CN [11]. By this NH_3-treatment,

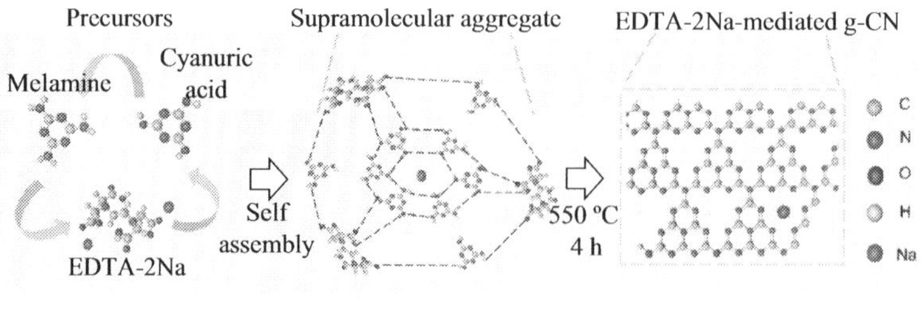

FIGURE 2.7 Derivation of supramolecular aggregates from melamine with cyanuric acid and EDTA-2Na [12].

S_{BET} increased from 0.32 to 135 m^2/g, while interlayer spacing d_{002} of *g*-CN remained unchanged. It exhibited a photocatalytic activity for CO_2 conversion.

g-CNs were prepared from melamine, supramolecular aggregates derived from a mixture of melamine with CA, as well as the EDTA-2Na mediated supramolecular aggregates by heating at 550°C for 4h in air, being coded as CN, SCN, and ESCN, respectively [12]. The synthesis process is illustrated in Figure 2.7. The CN was granular with a large particle size. A self-assembled supramolecular precursor with a sheet-like morphology with a large lateral size was formed through hydrogen bonds between the amino group on melamine and the hydroxyl group on CA. After calcination at 550°C, the release of H_2O and NH_3 from the supramolecules resulted in shrinkage and thinning, particularly ESCN being obtained as ultrathin flakes. The resultant *g*-CNs were tested for photocatalytic degradation of tetracycline hydrochloride (TC, an antibiotic) in water.

p-CNs were synthesized from the mixtures of melamine with either CA or trithiocyanuric acid (TCA), using aprotic DMSO and protic water (H_2O) solvents with different precipitation temperatures of 30–150°C [13]. The mixtures are coded as MCA and MTCA, respectively. Co-precipitates of melamine and CA were obtained spontaneously by mixing equimolar solutions in DMSO at a temperature between 30°C and 150°C. The precipitates at 30°C consisted of the hexagonal nanosheets with a thickness of about 40nm, which were aggregated into 2–3 μm spherical particles (MCA-DMSO-30), while the precipitates at 150°C consisted of spherical particles of closely packed arrays of hexagonal nanorods with a thickness of 200nm (MCA-DMSO-150). The crystals were heat-treated at a temperature of 500°C–600°C in N_2 to convert them into tri-*s*-triazine-based *g*-CN. Using H_2O solvent (MCA-H_2O at 30°C), the precipitates obtained were only bundles of thin microfibers with poor crystallinity. These bundles lost their preformed morphology during heat-treatment for polycondensation. The XRD patterns shown in Figure 2.8a–c reveal that the resulting *g*-CNs have similar crystallinity, despite the use of different precursors (melamine, CA, and TCA), the different solvents (DMSO and H_2O), and the different precipitation temperatures (30–150°C).

Highly crystalline *g*-CN was synthesized through the temperature-induced condensation of dicyandiamide ($C_2H_4N_4$, DCDA), using a eutectic mixture of LiCl/KCl (45/55 by mass, T_m = 352°C) as the solvent [14]. As shown in Figure 2.9, the precursor cyanamide polymerizes through deamination to form aromatic units and produces discrete oligomers, such as melam, melem, and melom. XRD patterns of

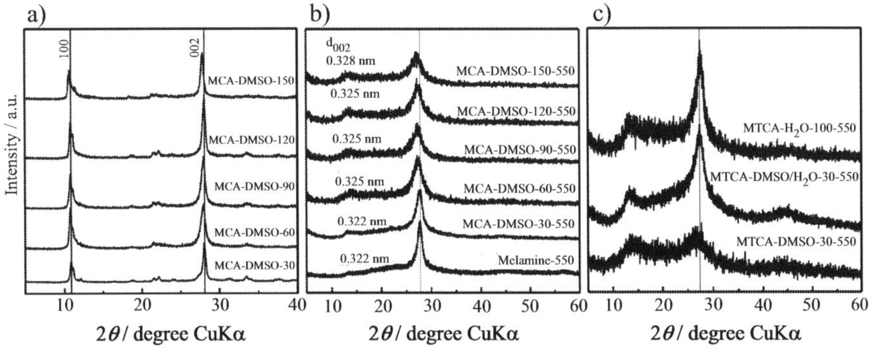

FIGURE 2.8 XRD patterns of the g-CNs prepared from the mixture of melamine with cyanuric acid (MCA) and trithiocyanuric acid (MTCA) precipitated from their DMSO and H_2O solutions at different temperatures (30–150°C) and heat-treated at 550°C: (a) effect of precipitation temperature, (b) after the heat-treatment at 550°C on the MCA-DMSO system, and (c) effect of solvents on MTCA [13].

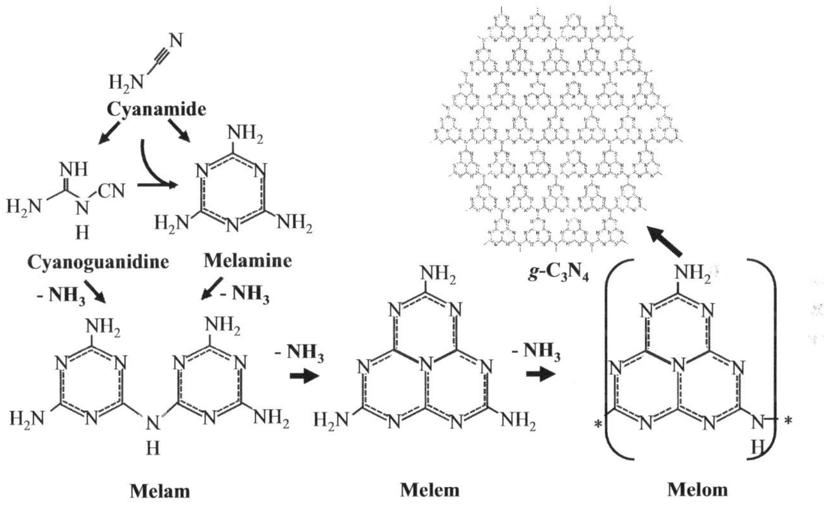

FIGURE 2.9 Structure changes from precursor cyanamide to g-CN [14].

the resulting g-CNs are shown in Figure 2.10a together with molecular structures. The lattice parameters of the resulting highly crystalline g-CN were determined as $a_0 = 0.8434$ nm and $c_0 = 0.6722$ nm. The synthesis of the crystalline g-CN (c-CN) has predominantly been achieved by the molten salt method, complemented by other approaches, such as solvothermal, template, self-assembly, and microwave-assisted methods [15,16]. Various molten salts in addition to KCl/LiCl have been employed, including KCl/NaCl and $KCl/ZnCl_2$ [17], KCl [18], NaCl/LiCl [19], $FeCl_3$/KCl [20], and Na_2CO_3 and Li_2CO_3 [21], among others. However, it must be pointed out that

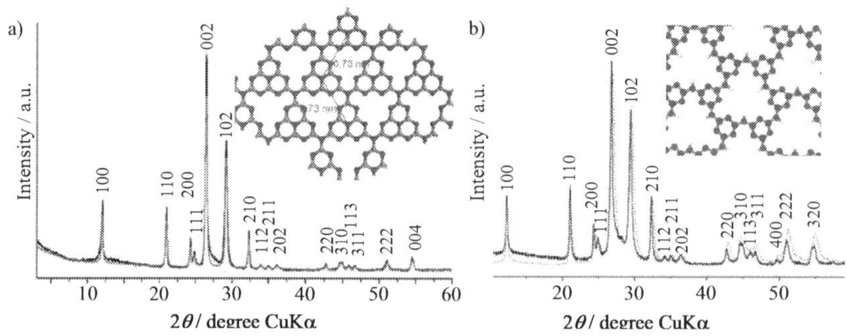

FIGURE 2.10 XRD patterns of the CNs synthesized using LiCl/KCl eutectic molten salt: (a) PHI [14] and (b) PTI with molecular structure of the respective CN (insets) [30].

the syntheses using molten salts did not always produce highly crystalline *g*-CN, although most of literature reports the synthesis of "crystalline *g*-CN". For example, the *g*-CN obtained using KCl/LiCl molten salt exhibited high crystallinity, showing sharp 002 in addition to some three-dimensional diffraction peaks (Figure 2.10a). In contrast, the *g*-CNs obtained by using KCl/NaCl and KCl/$ZnCl_2$ give broad 002 and 10 peaks but lack three-dimensional diffraction peaks as 102 and 210. This is similar to the *g*-CN synthesized using conventional calcination methods without using molten salt [17]. Different phenol degradation performances were observed among these *g*-CNs.

Crystallinity of *g*-CN could be improved with the help of a template, pressurization, and self-assembly. Polydimethylsiloxane (PDMS) mixed into the *g*-CN precursor worked as a template (soft template) during the polymerization of *g*-CN into highly crystalline nanosheets [22]. A Ni-foam (hard template) was immersed in an aqueous dispersion of DCDA at 80°C for 12 h, where the recrystallization of DCDA was anchored on the Ni foam, and followed by heating at 550°C for 2 h [23]. The resultant *g*-CN photocatalyst possessed high crystallinity with few structural defects, which could accelerate the rate of charge transfer and deliver enhanced H_2-evolution performance. Porous crystalline *g*-CN photocatalysts were synthesized by using a $NaHCO_3$ template [24]. 2 g DCDA was mixed with *X* g $NaHCO_3$ (*X* corresponds to 0, 0.05, 0.1, 0.15, 0.2, and 0.3) in deionized water and then freeze-dried to produce a $(NaHCO_3)_X$/DCDA mixture, followed by calcination at 550°C for 4 h to produce the porous crystalline *g*-CN. In this process, the $NaHCO_3$ worked first as the gaseous template to induce the formation of fluffy and porous pre-polymerized DCDA. The resulting Na_2CO_3 could further serve as the high-temperature solvent to promote the effective crystallization of the *g*-CN frameworks. The resultant porous crystalline *g*-CN delivered a significant increase in H_2-evolution rate. The calcination at 600°C for 2 h was repeated 7 times on the pristine *g*-CN synthesized by the calcination of DCDA at 550°C for 3 h [25]. This repeated calcination resulted in the increases in S_{BET} from 9.1 to 60.5 m²/g, the NO-oxidation efficiency from 20.6% to 35.8%, and the H_2-evolution rate from 31.3 to 764.8 μmol/g·h, while an improvement in XRD data was only slight, the 002 diffraction peak shifted from 27.26° to 27.58° in CuKα (a slight decrease in interlayer spacing). The effect of pressurization during the calcination of DCDA was studied by the calcination at 550°C in either a 50mL closed stainless steel reactor or an open reactor [26]. The g-CN obtained in a closed reactor exhibited a sharp 002 diffraction peak at 27.7° (d_{002} of 0.322nm), whereas the g-CN obtained in the open reactor gave 27.3° (0.327nm). A urea pellet compressed under 24MPa was heated by microwave (700 W power) for 30min [27]. The pellet exhibited an H_2-evolution rate of 61.7μmol/h, much higher than 10.6μmol/h for the g-CN derived from DCDA by calcination at 550°C in air.

Polymeric carbon nitride (p-CN) consists primarily of tri-s-triazine units, connected via secondary amines to form one-dimensional polymer chains. Hydrogen bonding interactions between these chains result in two-dimensional planes (g-CN framework). It is commonly synthesized by direct thermal polymerization of nitrogen-rich precursors at temperatures ranging from 450°C to 600°C, as shown in Figure 2.1. In contrast, the polymeric carbon nitrides composed of heptazine imide and triazine imide, i.e., poly(heptazine imide) (PHI) and poly(triazine imide) (PTI), respectively, are synthesized using the molten salt of LiCl and KCl, respectively. By heating a mixture of the precursor (i.e., melamine, DCDA, cyanamide, or urea) with a eutectic salt LiCl/KCl in a quartz ampoule under a closed vacuum system produced highly crystalline PTI sheets [28–30]. The XRD patterns of the PHI and PTI synthesized using the molten salt under a semi-closed and a closed vacuum system are compared in Figure 2.10a and b, respectively. The CNs synthesized using the eutectic molten salt (EMS) under different conditions have different atomic structures, as illustrated in their respective figure as an inset. The former (PHI) is composed of heptazine, while the latter (PTI) is composed of single triazine connected by a single nitrogen bond, as shown as an inset of their respective figures. During the PTI formation process, g-CN was formed as an intermediate and reported that a sealing of the reactor was essential for preventing an escape of the intermediates. A mixture of the DCDA-derived g-CN with CsCl (melting point of 645°C) in a 1/2 mass ratio was heated at 550°C for 2h in N_2 [31]. The resultant g-CN was a highly crystalline PHI nanosheet, suggesting that the unmolten CsCl crystals worked as a template for the growth of the PHI nanosheets. A recent review covered the structure, synthesis conditions, and electronic and photocatalytic properties of PHI and PTI [32].

High-quality g-CN with reduced structural defects was synthesized using microwave-assisted thermolysis of melamine-CA supramolecular aggregates [33]. In contrast to the conventional heating in an electric furnace, this approach enables the production of g-CN with much better crystallinity within minutes. The supramolecular aggregates prepared from a mixture of melamine and CA in equal mass (MCA complex) in DMSO were irradiated with microwave (700 W power), of which the temperature increased to approximately 1,000°C within less than 7min and kept for different periods of time (8–16min) to obtain CNs (coded as CN-X, where X corresponds to microwave irradiation period). For comparison, CN was synthesized from MCA by a conventional furnace heating at 540°C for 4h in Ar (coded as CN-540). The CN obtained by 10min microwave irradiation (CN-10) exhibited two distinct XRD peaks at $2\theta = 13.8°$ and 27.6° corresponding to 100 and 002, respectively. The intensity of the absorption in the visible light region increased in proportion to the duration of microwave irradiation period, from which bandgap was calculated to be 2.7eV for CN-16. Microwave irradiation was demonstrated to be a successful method to synthesize the sp^3-hybrid carbon nitride nano/micro-crystals [34]. A thin film of *g*-CN was synthesized by chemical vapor deposition (CVD) of melamine vapor [35]. The CVD deposition of *g*-CN was performed on an indium tin oxide (ITO) up to a film thickness of approximately 180 nm under the optimized conditions. However, the XRD pattern of the resultant *g*-CN film demonstrated only 002 and unsymmetrical 10 diffraction peaks, as observed by conventional calcination of melamine powder, poorer crystallinity than the one synthesized using EMS as previously mentioned [14]. In contrast, the CVD method was successfully employed to obtain a high-quality graphene sheet. A similar CVD process of melamine vaporized at 300°C and deposition onto various substrates, such as glass, silicon, or copper, at 550°C under N_2 flow in a tubular furnace also could not get crystalline films [36]. The resultant *g*-CN film appeared highly homogeneous and crack-free over a large surface area: the homogeneity was so high that it was difficult to determine whether the film had been deposited or not [37].

CN was regenerated from a solution of *g*-CN with $ZnCl_2$ molten salt hydrates, which was obtained by agitation of a urea-derived *g*-CN mixed with $ZnCl_2$ in water by heating to 120°C for 30 min, and by diluting with ethanol [38]. The resultant regenerated CN (*r*-CN) had a nanobelt morphology (about 20 nm width and hundreds of nm length), *C/N* atomic ratio of 0.609, and bandgap of 2.69 eV, while the pristine *g*-CN consisted of irregular sheets with considerable thickness, with a *C/N* atomic ratio of 0.615 and bandgap of 2.55 eV. *r*-CN gave a relatively sharp 002 diffraction peak, while no 10 diffraction peak. It delivered much higher H_2O_2-production performance than the pristine *g*-CN.

A carbon nitride with a composition $C_6N_9H_4Cl$ was synthesized through a solid-state reaction at 1.0–1.5 GPa and 500–550°C [39], consisting of the hypothetical graphitic phase of C_3N_4 (a PTI framework $C_6N_9H_3$) with approximately one HCl component for each unit of the framework. The Cl^--ion is located in the cavity V_{C3N} of the PTI framework, while the equivalent H^+ ion protonates the PTI (C–N) framework. Therefore, the compound may be expressed as $C_6N_9H_3{\cdot}HCl$. The $C_6N_9H_3{\cdot}LiCl$ was synthesized under ambient atmosphere [29]. A mixture of DCDA with a eutectic mixture of LiCl and KCl was heated in an Ar flow at 400°C for 12 h and then heated at 600°C for 48 h after sealing under evacuation into a silica glass tube to obtain a PTI intercalated by Li^+ and Cl^-. The elemental composition was measured as C 29.6, H 1.3, N 50.4, Cl 11.0, and Li 4.6 wt%. Its lattice parameters were determined as a_0=0.84684 nm and c_0=0.67502 nm, a slight increase in the parameter c_0, because Li^+ is located in the cavity V_{6N} and Cl^- is intercalated in the PTI layers, in other words, Li and Cl^--intercalated PTI. The Li and Cl^--intercalated PTI was successfully synthesized as single crystals [40]. Detailed studies on this single crystal suggested that the prismatic $\{10\bar{1}0\}$ planes had higher photocatalytic activity than the basal {0001} planes.

A mixture of DCDA with a eutectic mixture of LiBr and KBr (52/48 by mass, T_m of 348°C) in a weight ratio of 1/5 was heated in an inert atmosphere at 600°C for 48 h to get Br-intercalated CN [41]. The product kept a hexagonal unit cell with the lattice parameters of a_0=0.8500 nm and c_0=0.7045 nm (interlayer spacing d_{002} of 0.352 nm), a slight increase in c_0 being due to the Br-intercalation. The resultant Br-intercalated CN was heated at 60°C for 48 h in aqueous NH_4F solution (8 M) to get F-intercalated

CN, of which lattice parameters were determined as a_0=0.8421 and c_0=0.6638 nm, d_{002} being 0.332, smaller than that of graphite (d_{002} of 0.3354 nm).

The synthesis using the molten salt KCl/NaCl (1/1 by mass) resulted in the K (potassium) and Na (sodium) co-doped CNs [42]. Mixtures of the eutectic salts with melamine in 15/1, 30/1, and 45/1 by mass were heated at 520°C for 2 h and coded as PS_X-CN, where X corresponds to the mixing ratio of the salt (15, 30, and 45). The conduction band (CB) and valence band (VB) potentials of the resultant PS_X-CN were tunable from −1.09 and +1.55 eV, respectively, to −0.29 and +2.25 eV by controlling the mixing ratio of the eutectic salt. It exhibited high Rhodamine B degradation performances.

2.2 MOLECULAR MODIFICATIONS

The ionothermal approach on melamine using KCl/LiCl (EMS) as a high-temperature solvent yielded PTI [43], whereas the conventional thermal polycondensation resulted in a bulk g-CN (b-CN) based on heptazine. By changing the mass ratio of melamine/molten salt, the molecular units of CN from heptazines to triazines can be controlled [44], as illustrated in Figure 2.11. The melamine/EMS ratio of1/10 alters nearly all the molecular units of CN from heptazines to triazines. The reduction in the mixing ratio of melamine/EMS, such as 1/2, led to an incomplete transformation, and thus a molecular heterostructure of heptazine and triazine motifs, such as CN-2 in Figure 2.11, could be obtained. The product CN-2 contains a large amount of homojunctions, consequently exhibiting a marked improvement in photocatalytic activity for H_2O_2 production [44].

Two stable C-rich buckled CN nanomaterials based on pyrrole and imidazole frameworks, namely C_7N_6 and C_6N_6, were predicted using the first-principles

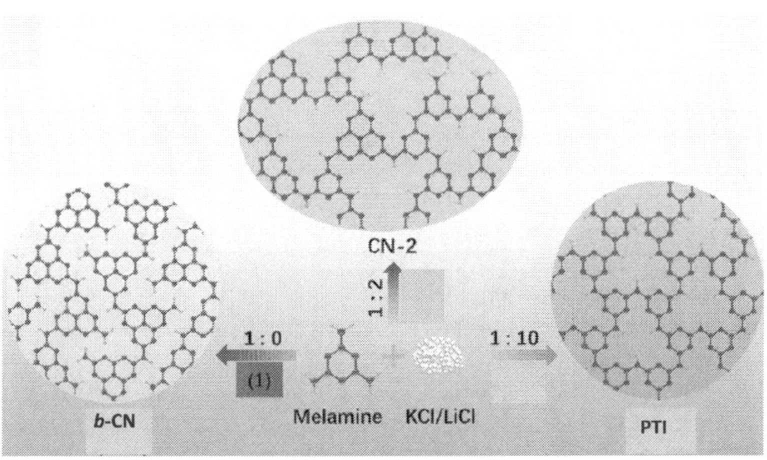

FIGURE 2.11 Schematic illustration of the synthesis of a molecular heterostructure of CN-2 using a melamine/EMS ratio of 1/2, and the synthesis of b-CN via conventional thermal polymerization using a melamine/EMS ratio of 1/0 and the synthesis of PTI using a melamine/EMS ratio of 1/10 [44].

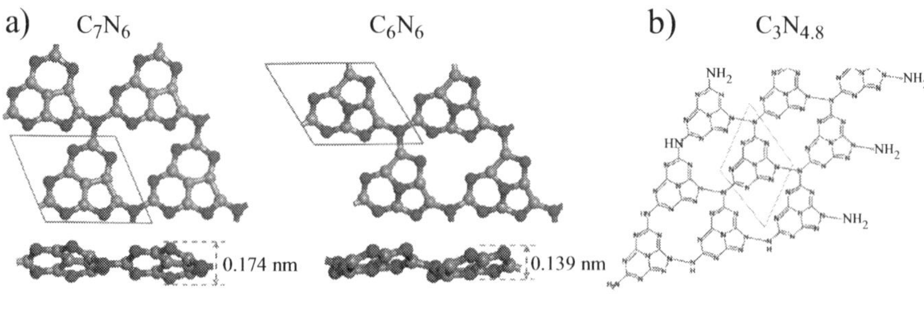

FIGURE 2.12 Atomic configurations for C- and N-rich CNs: (a) C-rich CNs, C_7N_6, and C_6N_6 (top and side view) and (b) N-rich CN, $C_3N_{4.8}$ [45,46].

TABLE 2.4
***C/N* Molar Ratio and BET Surface Area S_{BET}, Bandgap, and H_2-Evolution Rate of BA_X-CN [47]**

Catalyst	*C/N* Molar Ratio	S_{BET} (m^2/g)	Bandgap (eV)	H_2-Evolution Rate (μmol/h)
g-CN	0.73	10	2.67	148.2
$BA_{0.05}$-CN	0.74	22	2.57	253.1
$BA_{0.1}$-CN	0.75	19	2.45	218.8
$BA_{0.2}$-CN	0.76	14	2.26	175.7
$BA_{0.5}$-CN	0.80	12	1.86	115.3
$BA_{1.0}$-CN	0.88	11	1.68	53.9
$BA_{2.0}$-CN	0.96	12	1.58	8.5

computations [45]. Both C_7N_6 and C_6N_6 exhibited reduced bandgaps and broad optical absorptions in the visible region, comparable to those of the planar *g*-C_3N_4. In contrast, N-rich CN, $C_3N_{4.8}$, which was based on triazole and triazine moieties, was experimentally synthesized and its optical and electronic properties were discussed theoretically [46]. $C_3N_{4.8}$ was synthesized via a simple self-assembly of 5-amino-1H-tetrazole (5-ATTZ) as a new carbon nitride precursor using the composites of mesoporous silica with graphene (KIT-6-graphene hybrids). Atomic configuration for C-rich CNs, C_7N_6 and C_6N_6, and N-rich CN, $C_3N_{4.8}$, is shown in Figure 2.12a and b, respectively. A composite of the mesoporous $C_3N_{4.8}$ with a graphene sheet was found to be highly active for oxygen reduction reaction (ORR), exhibiting a higher diffusion-limiting current density and a decreased overpotential than those of *b*-C_3N_4 [46].

p-CN with different *C/N* atomic ratios was synthesized via co-polymerization of the mixtures of DCDA with barbituric acid ($C_4H_4N_2O_3$, BA) in different mixing ratios (0.05, 0.1, 0.2, 0.5, 1, and 2) in aqueous solution, followed by heat-treatment at 550°C (coded by BA_X-CN, where *X* represents BA mixing ratio) [47]. Bandgap of the catalysts and S_{BET} depend strongly on BA mixing ratio, i.e., *C/N* ratio, and as a consequence, H_2-evolution rate also, as shown in Table 2.4. The BA-mixing ratio of 0.05 (the catalyst $BA_{0.05}$-CN) exhibited the highest H_2-evolution rate.

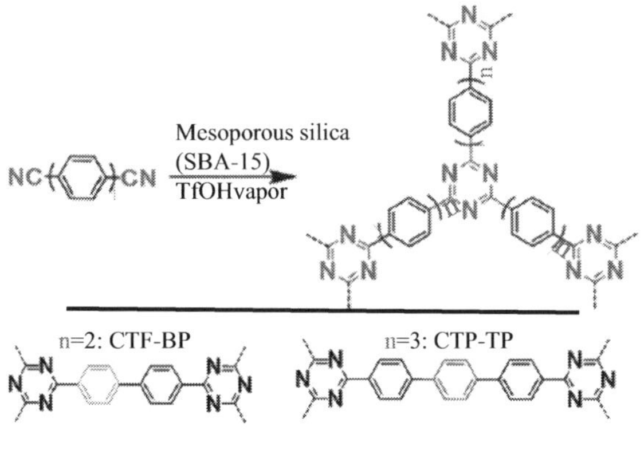

FIGURE 2.13 The synthesis procedure and structures of CTP-BP and CTP-TP units [48].

The covalent triazine polymers (CTPs) containing biphenyl (BP) and terphenyl (TP) units were prepared via cyano-polymerization in the presence of trifluoromethyl sulfonic acid (TfOH) vapor using mesoporous silica (SBA-15) as the carrier, the synthesis procedure is shown in Figure 2.13 [48]. The molecular modification of the triazine polymer to CTP-BP and CTP-TP effectively enhanced the H_2O_2 production.

During thermal polycondensation, the 2D hexagonal assembly between melamine and CA evolved into nanosheet-like structures without losing its macroscopic morphology. CA and other triazines, such as TCA, are known to form hydrogen-bonded networks that give rise to different structures depending on the donor–acceptor pair and crystallization solvent [49]. g-CN was synthesized by selecting various donor–acceptor pairs, such as melamine with CA and TCA (MCA and MTCA, respectively), using different solvents, aprotic DMSO and protic H_2O, and different precipitation temperatures, 30–150°C [13]. Co-crystals of MCA were spontaneously precipitated by mixing equimolar solutions in DMSO at 30°C (coded as MCA-DMSO-30), which were composed of 2–3 μm spherical aggregates of the hexagonal nanoplates with a thickness of about 40 nm. Their morphology changed from hexagonal nanoplates to hexagonal nanorods with increasing precipitation temperature to 150°C. Close-packed arrays of hexagonal nanorods with a thickness of 200 nm were observed in the spherical particles of MCA-DMSO-150. MCA-DMSO could be transformed to the triazine-based g-CN by heating at 550°C in N_2. Upon using H_2O solvent, the HT (180°C for 4 h) was employed to increase the solubility of MCA in H_2O. Precipitation of MCA in H_2O at 30°C (MCA-H_2O-30) yielded only bundles of thin microfibers with a diameter of 1 μm and a length of up to 10 μm, having poor crystallinity. The photocatalytic activity of the resultant macroscopic hydrogen-bonded networks (MCA and MTCA-derived g-CN) delivered an enhanced H_2-evolution performance [13].

The accurate tuning of photocatalytic activity was attempted through a modification of the molecular structure by loading metal nanoparticles onto its surface and/or doping heteroatoms into its structure (molecular modification) [50,51]. The melem, which was synthesized by calcination of melamine at 425°C, was mixed with phosphonitrilic chloride trimer (PCT) in a molar ratio of 1/1 and calcined at 500°C,

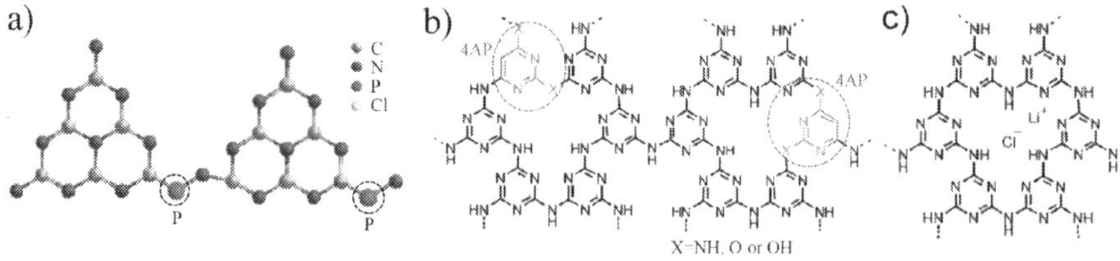

FIGURE 2.14 Molecular modification (a) by doping P [51], (b) by doping 4-amino-2,6-dihydroxypyrimidine (4AP), and (c) Li^+Cl^- [50].

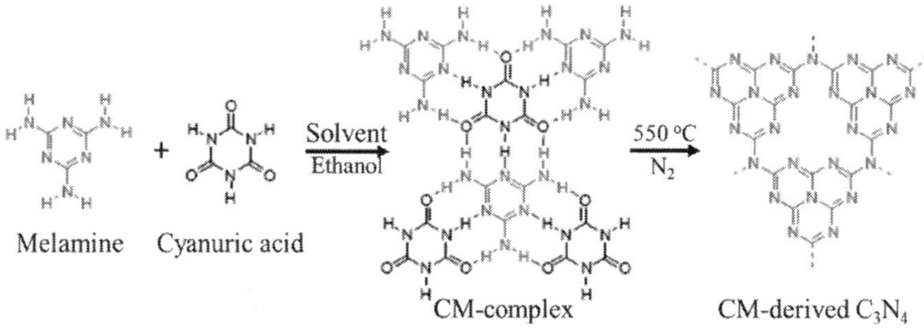

FIGURE 2.15 Syntheses of CM-complex and CM-derived CN [54].

550°C, 600°C, and 650°C for 4h to produce P-doped CN (P-CN) [51], as shown in Figure 2.14a. It was also modified using 4-amino-2,6-dihydroxypyrimidine (4AP) and Li^+Cl^- [50], as shown in Figure 2.14b and c, respectively. These modified *g*-CN delivered an enhanced H_2-evolution rate.

Three kinds of polyimide (PI) were obtained: the conjugation of the melem with a di-anhydride monomer, such as pyromellitic dianhydride (PMDA), 4,4-biphthalic dianhydride (BPDA), and naphthalene-1,4,5,8-tetracarboxylic dianhydride (NTDA) in equal molar ratio by heating in a semi-closed system at 350°C for 4h [52].

The CA-melamine complex (CM-complex) forms different ordered textures [53]. Its particles obtained from ethanol solution exhibited a highly ordered pancake-like morphology, leading to the formation of hydrogen-bonded supramolecular structure (CM-complex) and the conversion to *g*-CN by calcination at 550°C in N_2, as shown in Figure 2.15 [54]. The photocatalytic activity of the CM-derived CN was higher than that of the DCDA-derived CN, depending strongly on the solvent used, ethanol, water, chloroform, etc.

2.3 TEXTURE MODIFICATIONS

2.3.1 Templating

On carbon materials, various templating techniques have been developed using various templates, such as zeolite, SiO_2, and MgO, and applied to prepare porous carbons with controlled pore structures [55–57].

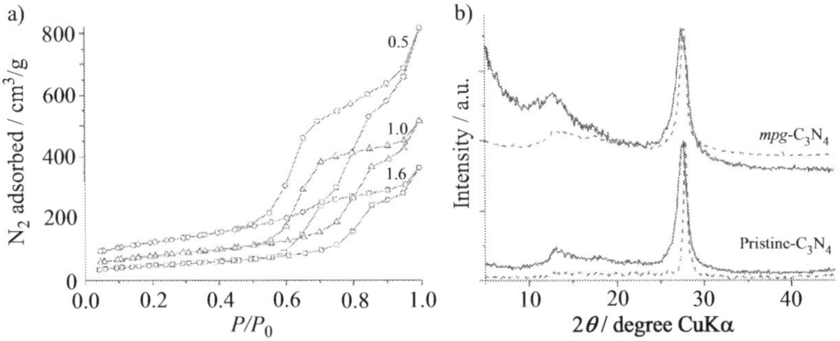

FIGURE 2.16 Syntheses of mesoporous CNs (*mpg*-CNs) using mesoporous silica template: (a) N_2 adsorption–desorption isotherms of *mpg*-CN synthesized using different amounts of template and (b) XRD patterns [58].

TABLE 2.5
H_2-Evolution Rate of *mpg*-CN Loaded with Pt (3 wt%) [59]

Photocatalyst	Silica/CA by Mass	S_{BET} (m^2/g)	H_2-Evolution Rate (μmol/g·h)
g-C_3N_4	–	8	18
mpg-C_3N_4	0.2	67	149
	0.5	126	142
	1.0	235	124
	1.5	373	69
g-C_3N_4	NH_4NF_3-washed	10	19

By dispersing a precursor cyanamide (CA) with mesoporous silica (hard template) in a mass ratio of 0.5, 1.0, and 1.6 and following heat-treatment and washing with NH_4NF_3 to remove SiO_2, mesoporous g-CN (*mpg*-CN) was synthesized [58]. It exhibited a marked enhancement of H_2-evolution performance. The formation of mesopores is confirmed from the hysteresis of the N_2 adsorption–desorption isotherm as shown in Figure 2.16a. From XRD patterns (Figure 2.16b) of the *mpg*-C_3N_4, its crystallinity was improved by the heat-treatment at 600°C for 10 h. The H_2-evolution rate of the Pt-loaded (3 wt%) *mpg*-C_3N_4 catalysts is listed in Table 2.5, revealing that texture modification by introducing mesopores is effective to enhance activity [59]; particularly, the H_2 evolution was accelerated markedly using a small mixing ratio of 0.2. Washing with NH_4NF_3 to eliminate the silica template does not have any effect on photoactivity. Silica-templating was also performed by mixing CA with tetraethyl orthosilicate (TEOS) in a TEOS/CA mass ratio of 1/3, 1/6, and 1/12 (coded as CN/SH-3, -6, and -12) [60]. Pore structure parameters are compared with the resultant composites of CN with SiO_2 (before washing to remove SiO_2), the resultant *mpg*-CN (after washing), and the mesoporous silica obtained by heating the composite at 650°C in air to remove CN (SH) in Table 2.6, revealing the formation of *mpg*-CNs.

TABLE 2.6
Pore Structures of the Porous CN (CN after Silica Removal), Mesoporous Silica (SH), and Their Composite CN/SH (before Silica Removal) [60]

	Code	TEOS/CA Molar Ratio	S_{BET} (m²/g)	V_{total} (cm³/g)	V_{micro} (cm³/g)	V_{meso} (cm³/g)	Average Pore Diameter (nm)
Before washing (CN/SH composite)	CN/SH-3	1/3	270	0.31	0.03	0.28	5.6
	CN/SH-6	1/6	77	0.16	0.01	0.15	16.4
	CN/SH-12	1/12	6	0.01	–	0.01	–
After washing out SiO_2 (porous CN)	CN-3	1/3	1	0.01	0.01	–	–
	CN-6	1/6	131	0.16	0.05	0.11	2.4
	CN-12	1/12	167	0.19	0.02	0.17	3.6
Mesoporous silica (SH)	SH-3	1/3	864	0.56	0.48	0.08	2.6
	SH-6	1/6	939	0.92	0.46	0.46	4.9
	SH-12	1/12	901	1.17	0.52	0.65	6.5

2.3.2 Exfoliation

On graphite, exfoliation was employed to make graphite flakes thinner to obtain graphene flakes (thinning) and/or to obtain graphite foam (aerogels) either through thermal treatment of graphite intercalation compounds, often using sulfuric acid (thermal exfoliation), with microwave irradiation (microwave exfoliation), electrochemical treatment in a liquid electrolyte (electrochemical exfoliation) or in a surfactant solution, such as sodium dodecylsulfate (SDS) and sodium dodecylbenzene sulfonate (SDBS) (surfactant-assisted exfoliation) [61].

Exfoliation of *p*-CN (amorphous CN) into thin 2D nanosheets can lead to enhanced surface activity and unique optical and electronic properties [62]. Four exfoliation strategies (ultrasonication-assisted, thermal, hydrothermal, and electrochemical) were employed to understand the relationship between the exfoliation mechanism and the structural/optical properties of CNs. The thermal and ultrasonic exfoliation can effectively reduce the thickness of *p*-CN, preserving its original structure. In contrast, the chemical and hydrothermal exfoliations strongly influenced the morphology and structure of *p*-CNs, leading to a decreased performance for phenol photodegradation.

Bulk CN (*b*-CN) synthesized from melamine at 550°C for 4h was exfoliated via the intercalation of NO_2^- under hydrothermal condition (at 100°C for 12h) in diluted HNO_3 aqueous solution, followed by annealing at various temperatures of 450°C, 500°C, 530°C, and 570°C for 3h to get few-layered CNs (coded by *fl*-CN-450, -500, -530, and -570, respectively) with an average yield of 82.6%, 58.8%, 9.6%, and 7.2%, respectively [63]. From AFM analysis of the thickness of the sheets, the average number of layers stacked was measured to be about 95, 75, 53, 8, and 2 layers. The exfoliation into few-layered nanosheets was effective in enhancing H_2O_2 production. The exfoliation of *g*-CN strongly influences on the CB potential E_{CB}, increasing from −0.72eV for the pristine *b*-CN to −0.86 and −0.92V for the *fl*-CN-530 and 570,

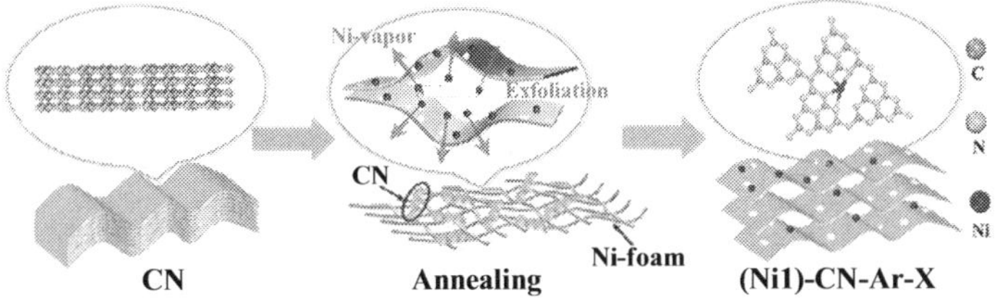

FIGURE 2.17 Schematic illustration of the synthesis procedure of the Ni-intercalated g-CN catalysts((Ni1)/CN-Ar-X) via metal-vapor exfoliation procedure [64].

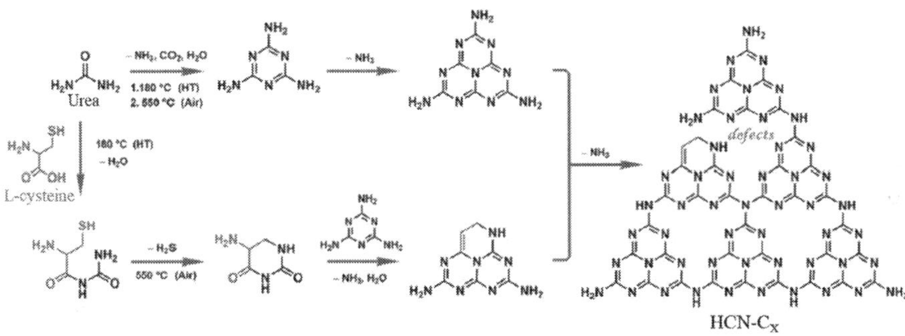

FIGURE 2.18 Schematic illustration of the synthesis route to HCN-C_X [65].

respectively, while the VB potential E_{VB} was a constant at around 2.02 eV, in other words, bandgaps increased by exfoliation from 2.74 eV for the pristine to 2.87 and 2.93 eV after exfoliation at 530°C and 570°C, respectively.

p-CN powder synthesized from melamine at 550°C for 2 h in air was spread evenly on nickel foam, of which the two pieces were pressed together and heated in an alumina crucible with a cover at 550°C for different times in Ar to get single Ni-atom (Ni1) intercalated g-CN (coded as (Ni1)-CN-Ar-$\boldsymbol{X}$, where X corresponds to holding time at 550°C [64]. This process, employed for the preparation of the photocatalyst, was called metal-vapor exfoliation. The exfoliation processes and the resultant structures, (Ni1)-CN-Ar-X, are illustrated in Figure 2.17. The exfoliation was effective in enhancing the performances for the CO_2-reduction to CO and CH_4.

2.3.3 Doping

C-rich CN nanosheets were synthesized by mixing urea (0.5 mol) with L-cysteine ($C_3H_7NO_2S$) of 0.0, 0.1, 0.3, 0.5, and 1.0 mmol in deionized water, dehydrating at 100°C, and then heating at 180°C in an autoclave, followed by the calcining at 550°C for 2 h in air and coded as HCN-C_X, where X was L-cysteine used during the synthesis (0.0–1.0 mol) [65]. The synthesis route of HCN-C_X is schematically illustrated in Figure 2.18. The introduction of –C=C–C skeleton (highlighted by light grey in Figure 2.18) from L-cysteine could modify the band structure of CN and its bandgap

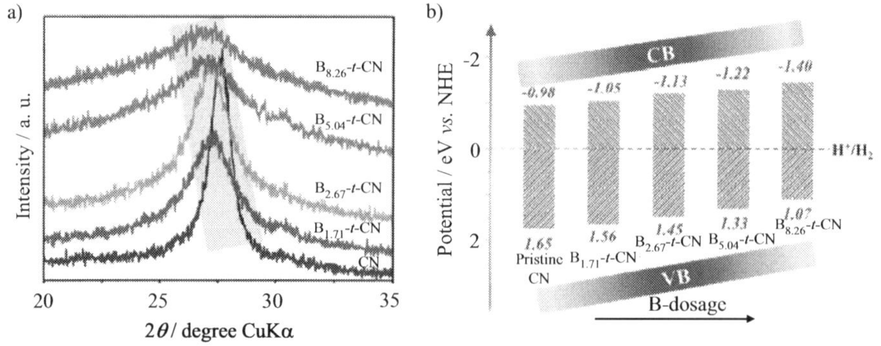

FIGURE 2.19 B-doped CN tubes (B_X-*t*-CN): (a) XRD patterns and (b) bandgaps [66].

becomes smaller, depending on the dosage X, from 2.68 eV for the pristine *g*-CN to 2.49 eV for the HCN-$C_{1.0}$.

B-doping into *g*-CN was possible, as into carbons, by mixing boric acid with *g*-CN precursor (melamine and CA), followed by heating at 550°C [66]. B-doped *g*-CNs with different B-dosages and tubular particle morphology were synthesized by using different mixing ratios, of which the B-content was determined as 1.71, 2.67. 5.04, and 8.26 at%, respectively, coded as B_X-*t*-CN, where *X* corresponds to B-dosage (1.71–8.26), respectively. The resultant $B_{2.67}$-CNT, for example, consisted of tubes with a diameter and length of 300–500 nm and 2–3 μm, respectively, with many macropores on their walls. The added H_3BO_3 facilitated B-doping, forming macropores by maintaining tubular morphology. In Figure 2.19a, 002 diffraction profiles of B_X-*t*-CNs are compared, revealing that the interlayer spacing increased with increasing B-dosage, probably due to the deformation of the lamellar structure by the B-doping. In Figure 2.19b, the bandgaps calculated on the basis of UV-vis spectra are shown, revealing that the bandgaps of B-CNTs progressively decreased from 2.63 to 2.47 eV. This suggests that the light absorption range of B-CNTs gradually increase.

In B-doped CN, the replacement of a C-atom with a B-atom is possible in three positions in *g*-CN, as shown in Figure 2.20a and b, respectively [67]. The first position is bonding with 2 N-atoms and is expressed as B_{2N}, the second position is bonding with 3 N-atoms (expressed as B_{3N}), while the third position is bonding with 3 C-atoms (B_{3C}). The theoretical DFT calculation predicted that the third position of B (i.e., B_{3C}) is the active site for the performance of photocatalytic reactions, including N_2-fixation.

B-doped *g*-CN was synthesized by calcining a mixture of *g*-CN with different amounts of BH_3NH_3 at 600°C in N_2 flow and coded as B_X-CN, where *X* is at% of B-doped (0.5–3.0 at%) [68]. Photocatalytic performances for CO_2 reduction were electrochemically investigated using the resultant B_X-CN catalysts as the cathode, demonstrating the preferential formation of C_2H_5OH.

The single-atom Co- and Mn-metals (expressed as Mn1 and Co1, respectively) co-doped CN was synthesized via a supramolecular self-assembly [69]. First, an equal molar ratio of TCA and melamine with an appropriate amount of $Mn(CH_3COOH)_2$

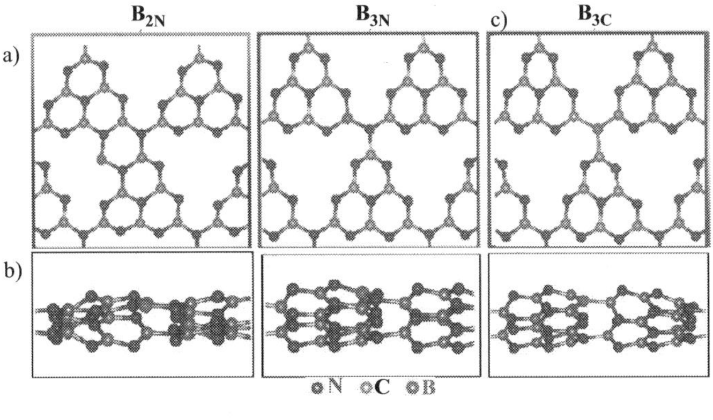

FIGURE 2.20 Possible three positions for doped B in g-CN molecule, B_{2N}, $B_{3N,}$ and B_{3C}: (a) top views, (b) side views and (c) NRR active site [67].

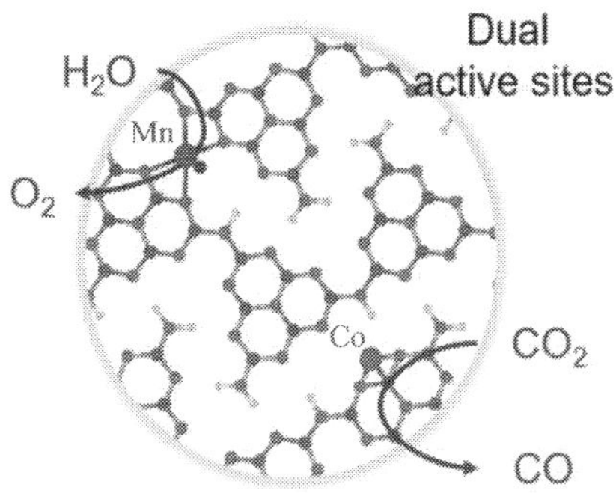

FIGURE 2.21 Structure of (Col), (Mnl)-CN by highlighting the single-atom active site with functions [69].

were stirred in water at a low speed to prepare supramolecular adducts, which were calcined at 550°C in Ar to get (Mn1)-doped CN microrods. The resultant (Mn1)-doped CN was dispersed in an aqueous solution of $Co(NO_3)_2$ and then annealed at 350°C in H_2 to get Mn- and Co-co-doped CN (coded as (Mn1),(Co1)-CN). The resultant (Co1), (Mn1)/CN has dual single-atom active sites for simultaneous H_2O oxidation and CO_2 reduction, as illustrated in Figure 2.21.

2.4 g-C_3N_4-BASED COMPOSITES

2.4.1 Self-Composite

Construction of an isotype heterojunction between two different crystal phases of a single substance (i.e., crystal-phase heterojunctions or homojunctions) could enhance the photocatalytic activity, as in the case of a commercially available TiO_2 (P25, a mixture of 75% anatase-TiO_2 and 25% rutile-TiO_2) [70]. A similar crystal-phase heterojunction was formed in *g*-CN prepared by thermal condensation of dicyandiamide ($C_2H_4N_4$, DCDA) as a single precursor [71]. DCDA was heat-treated at 600°C for 4 h in N_2 flow to obtain *g*-CN (CN-600). This CN-600 (0.5 g) was mixed with different amounts of the pristine DCDA (0.2, 0.4, and 0.6 g) in water and heated at 80°C to evaporate solvent water, and the resultant powder mixtures were subjected to the heat-treatment at 500°C to convert DCDA to *g*-CNs to get homojunctions between CN-600 and CN-500 (CN-500/CN-600 self-composite). With increasing the preparation temperature, crystallinity of *g*-CN decreased gradually, where the 10 diffraction peak in the XRD pattern became weaker and broader, and the 002 peak slightly shifted to a higher 2θ value. SEM image of CN-500 exhibited a pronounced aggregation along with discernible layered morphology, whereas CN-600 exhibited a compact and rather largely layered structure (a crystalline CN).

A self-composite $C_3N_4/C_3N_{5.4}$ was synthesized through a self-assembly and copolymerization of the mixture of melem with aminoguanidine hydrochloride (CH_6N_4·HCl) in different mass ratios (5/1, 5/2, 5/3, and 5/4) in water, followed by heating and stirring the solution until all the water evaporated. The resultant precipitates were calcined at 500°C in N_2 flow [72]. On these composites, the coexistence of the C_3N_4 and $C_3N_{5.4}$ moieties was confirmed by ^{13}C NMR and ^{1}H NMR measurements. The resultant composites exhibited much higher and faster adsorption and degradation of As(III) and As(V) in wastewater.

The self-composites of B-CN/CN were prepared from the mixtures of urea with B-doped CN (B-CN_X/CNs) exhibited improved degradation performances for methyl orange and phenol [73]. The composites B-CN_X/CNs were synthesized from the mixtures of 10 g urea with different amounts of B-CN (10, 25, and 50 mg) by calcining at 550°C for 2 h, to obtain the composites B-CN_{10}/CN, B-CN_{25}/CN, and B-CN_{50}/CN, respectively. B-CN was prepared by calcination of the mixture of 10 g urea with 6.0 mg sodium tetraphenylborate (Ph_4BNa) at 550°C for 2 h. Photoluminescence (PL) spectrum for the composite B-CN_{25}/CN was compared with the pristine CN and the B-CN at room temperature with an excitation wavelength of 400 nm. While the position of the emission peak for the composite B-CN_{25}/CN hardly changed in comparison with the pristine CN and B-CN, its relative intensity was lower. This suggests that the separation efficiency of the photogenerated electrons and holes can be improved by the band offsets in the B-CN/CN heterojunction. The composite B-CN_{25}/CN exhibited significantly higher degradation efficiencies for methyl orange and phenol, as well as a H_2-evolution rate, than the pristine CN, B-CN, and other B-CN_x/CNs (B-CN_{10}/CN and B-CN_{50}/CN).

A novel carbon nitride homojunction with the electron-enriched and hole-enriched CN nanosheets (coded as O_x-CN and Re-CN) was proposed by synthesizing the self-composite Re-CN/CN/O_x-CN [74]. The spray-calcination method was employed to synthesize CN nanosheets containing different concentrations of carbon vacancies. The melamine dispersed aqueous solution was sprayed onto the inside of a preheated crucible (120°C) by controlling the solid film content to be 4.83 g after water removal, on which a dense urea film was formed. Subsequently, 0.42 g thiourea and 0.75 g melamine were loaded sequentially by a similar method. Finally, the resultant layer-by-layer precursor film was calcined at 550°C for 2 h to produce Re-CN/CN/O_x-CN homojunction composite.

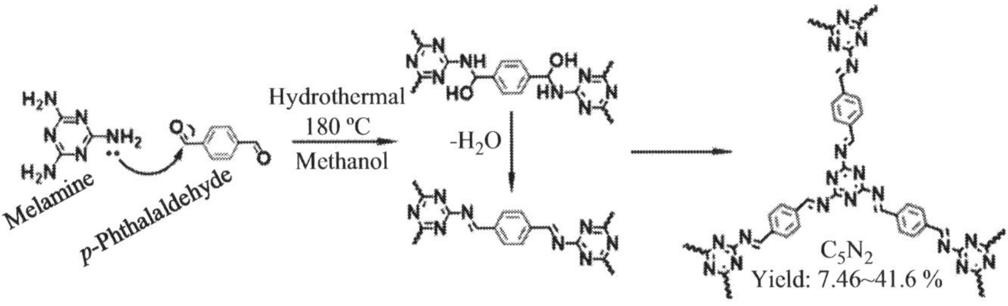

FIGURE 2.22 Synthesis procedure of C_5N_2 with a conjugated C=N linkage [76].

This composite exhibited efficient photocatalytic activity for H_2 and O_2 evolution (127.05 and 3.60 mmol/h·g, respectively) with a scavenger reagent.

A ternary carbon nitride homojunction was prepared by mixing amino-rich CN (A-CN), oxidized CN (O-CN), and phosphorylated CN (P-CN) in equivalent weight in water under sonication, centrifugation, washing with ethanol, and drying at 80°C [75]. The mixture was dispersed in a homogeneous aqueous solution of pectin and the resulting colloid solution was dropped into a 0.1 M $FeCl_3$ solution to form spherical hydrogels, followed by crosslinking for 24 h, and finally washed with water to get AOP-CN hydrogel. A-CN and O-CN were prepared from the mixture of melamine with oxalic acid, and melamine with ammonium oxalate, respectively, by calcining at 550°C. P-CN was prepared by mixing g-CN with sodium phosphate solution and stirring for 24 h and dried at 80°C. The resulting self-composite AOP-CN containing ternary homojunction was subjected to study H_2 evolution, H_2O_2 production, and tetracycline degradation.

A well-defined C_5N_2 with a conjugated C=N linkage was synthesized from a methanol solution of melamine with p-phthalaldehyde in a 2/3 mass ratio by HT at 180°C for different times (0.5–5 days), followed by Soxhlet extraction [76]. The synthesis procedure of C_5N_2 is illustrated in Figure 2.22. The resultant C_5N_2 catalyst exhibited a much higher H_2O_2-production rate (1,550 μmol/L·h) than that of the pristine g-CN (5.2 μmol/L·h).

2.4.2 With Metals

In addition to doping of single metal atoms in the intrinsic cavities of g-CN, nanoparticles of various metals were loaded onto the surface of g-CN through in situ photo-deposition in $(NH_4)_2RuCl_6$, Na_3RhCl_6, $(NH_4)_2PdCl_4$, Na_2IrCl_6, H_2PtCl_6, and $HAuCl_4$ as a co-catalyst for photocatalytic reactions. On Pt and Ru, loading was performed by an impregnation method using THF solutions of bis(1,5-cyclooctadiene) platinum complex, $Pt(cod)_2$, and triruthenium dodecacarbonyl, $Ru_3(CO)_{12}$. The metal loading enhanced the H_2-evolution rate, depending on the metal loaded: Pt-loading was the most effective, while Ir had only a little promotional effect [77].

K^+-loading into g-CN was effective to enhance H_2O_2 production [78]. K-loaded crystalline g-CN (K/CN) was synthesized from a mixture of 20 g of urea, varying amounts of KCl (0.5, 1, and 2 g) in an appropriate amount of ethanol by heating in a crucible sealed by tinfoil at 550°C for 2 h. The products were coded as K_X/CN, where

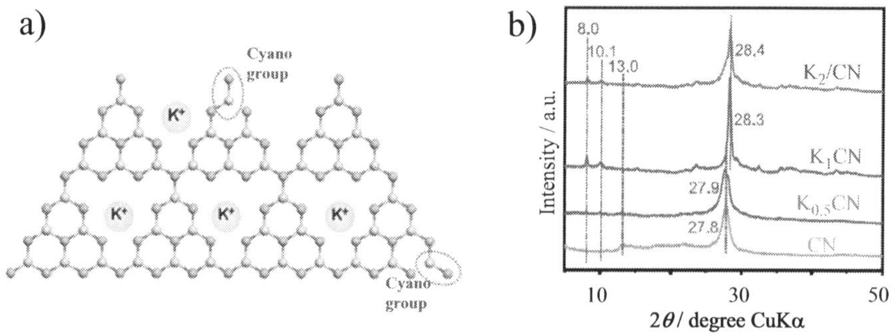

FIGURE 2.23 K-loaded g-C_3N_4 (K_X/CN): (a) illustration of K^+-location in g-CN and (b) XRD patterns of K_X/CNs [78].

X corresponds to the amount of KCl mixed into 20 g urea (0.5, 1, and 2 g). The molecular structure of K_X/CN and XRD patterns for K_X/CNs are shown in Figure 2.23a and b, respectively. It is reasonably supposed that K^+ is located in an intrinsic cavity V_{6N}, as illustrated in Figure 2.23a, this K^+ positioning causing no enlargement of the interlayer spacing of the matrix g-CN, as shown in Figure 2.23b.

Cu atoms were uniformly anchored on the surface of S-doped CN (S-CN) by dispersing S-CN in a Cu phytate (CuPA) dispersion, followed by microwave irradiation [79]. Various amounts of CuPA (X of 5, 10, 15, or 20 mg for 100 mg S-CN) were loaded with 800 W microwave irradiation. The resultant composites, coded as $(CuPA)_X$/S-CN, delivered enhanced CO_2-conversion performances, the $(CuPA)_{15}$/S-CN exhibiting the highest CH_3OH-yield reaching 11.87 μmol/g·h.

2.4.3 With Inorganics

p-CN was loaded by N-doped $K_2Ta_2O_6$ (KT) and $H_2Ta_2O_6$ (HT) by means of a solid-state reaction method with urea as the CN precursor [80]. The powder mixture of urea with $K_2Ta_2O_6$ in a 3/1 mass ratio was heated at 400°C for 2 h in a crucible with a lid to get N-doped KT-loaded CN (KT/CN). The resultant KT/CN composite was treated in aqueous HCl solution to get HT-loaded CN (HT/CN). Both composites KT/CN and HT/CN are photoactive, exhibiting a high H_2-evolution rate of 50 and 78 μmol/h·g, respectively.

NiO nanoparticles-loaded g-CN was wrapped with mono-layer reduced graphene oxide rGnO to produce a composite (rGnO/NiO/CN) [81]. The NiO/CN composite (NCN) was prepared by mixing NiO nanosheets with g-CN nanowires in a 2/1 weight ratio and heating at 300°C in N_2, which was mixed into a commercially available GnO aqueous dispersion (5 wt% concentration) at 120°C to get the composite. Different rGnO-contents (1, 3, 5, 7, and 10 wt%) were employed and the resultant composites were coded as G_X/N/CNs, where X corresponds to rGnO-content (1–10 wt%). The preparation procedure of the composite G_X/N/CN and their UV-vis spectra are shown in Figure 2.24a and b, respectively. These composites exhibited high photocatalytic CO_2-conversion performances with a high conversion to CO and CH_4, depending on rGnO-content of the composite, and stable cyclabilities.

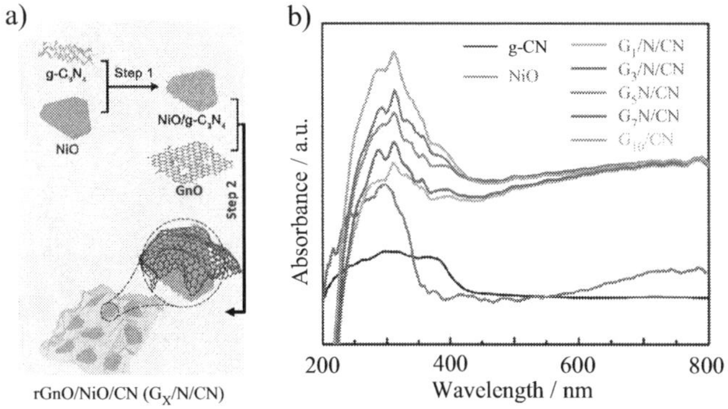

FIGURE 2.24 The rGnO-wrapped NiO/CN composites G_X/N/CNs: (a) preparation procedure and (b) UV-vis spectra [81].

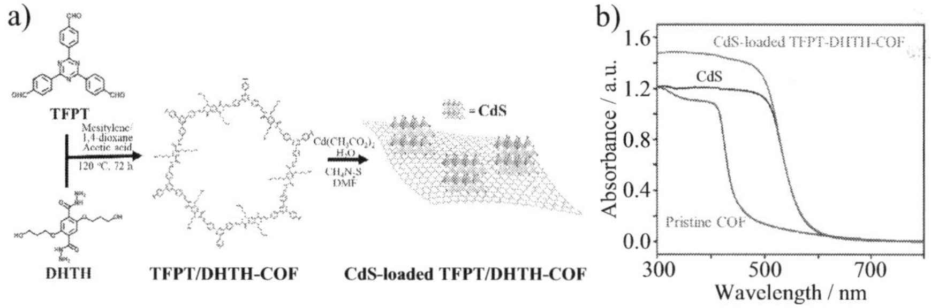

FIGURE 2.25 CdS-loaded TFPT/DHTH-COF: (a) Synthesis procedure and (b) UV-vis absorption spectrum in comparison with CdS and the pristine COF [82].

The triazine-based covalent organic framework (COF) was synthesized via a condensation reaction of precursors, 1,3,5-tris-(4-formyl-phenyl)triazine (TFPT) and 2,5-bis-(3-hydroxypropoxy) terephthalohydrazide (DHTH), coded as TFPT/DHTH-COF [82]. The resultant COF (10–30 mg) was dispersed in 10 mL deionized water containing 446 mg cadmium acetate dihydrate, in which 60 mL of DMF and 669 mg thiourea were added, and then hydrothermally heat-treated at 160°C for 12 h to produce CdS-loaded TFPT/DHTH-COF. The synthesis procedure is illustrated in Figure 2.25a. In Figure 2.25b, UV-vis spectra are shown, revealing that the bandgaps of the pristine COF of 2.77 eV become closer to the bandgap of CdS nanoparticles (2.34 eV), and CdS-loaded COF composite displays stronger light-harvesting capability compared to CdS nanoparticles.

Two kinds of MoS_2/g-CN composites were prepared, the one being by HT of an aqueous solution of the mixture of melamine-derived g-CN with MoS_2-precursors ($(NH_4)_6Mo_7O_{24}$, NH_2CSNH_2, and HCl) at 200°C for 24 h (coded as MoS_2/CN-HT) and another being by calcination of the mixture of g-CN with MoS_2 in equal mass at 300°C for 2 h in N_2 (coded as MoS_2/CN-Cal) [83]. The loading of MoS_2 onto the g-CN scaffold to create the MoS_2/CN heterojunctions significantly increases the light

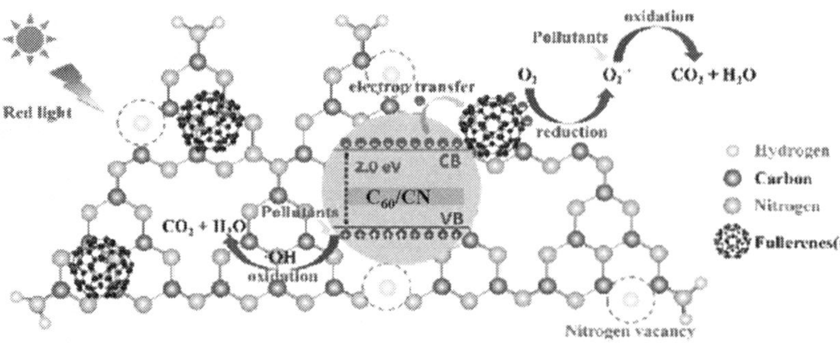

FIGURE 2.26 Illustration of C_{60}-modified *g*-CN (C_{60}/CN) and its functions for pollutant degradation [85].

absorption region compared to bulk MoS_2, leading to the accelerated production of photogenerated electrons and holes, due to the cooperative effect of MoS_2 and *g*-CN. The bandgaps for the composites MoS_2/CN-HT and MoS_2/CN-Cal were calculated to be 2.47 and 2.53 eV, respectively. The resultant composites delivered the enhanced degradation performances for Rhodamine B.

Loading of Ag-doped CdS nanoparticles onto the surface of S-doped *g*-CN (Ag-CdS/S-CN composite) was effective to improve the photocatalytic degradation performance, higher than the respective component catalysts (S-CN, CdS, Ag-CdS, and CdS/S-CN) [84]. The composite was prepared via the deposition of Ag-CdS nanoparticles on S-CN synthesized from thioacetamide at 600°C.

The *mpg*-CN modified with C_{60} molecules markedly accelerated rhodamine B (RhB) photodegradation rate [85]. C_{60}-modification of *mpg*-CN was performed by mixing C_{60} (0.1, 0.3, 0.5, 0.7, and 0.9 wt%) into an aqueous dispersion of melamine at 60°C for 10 h and followed by heating at 550°C for 4 h to get C_{60}-loaded CN (C_{60}/CN). The structure of the resultant C_{60}-modified *mpg*-CN is illustrated in Figure 2.26. C_{60}-modification enhanced the RhB-degradation rate by more than 25.8 times that of bulk *g*-CN and by 7.8 times that of the pristine *mpg*-CN.

The rGnO/CN composite aerogel was prepared [86]. The mixture of 0.4 g graphite oxide GtO with 0.2 g melamine-derived *g*-CN dispersing in deionized water was hydrothermally treated at 160°C for 12 h, of which the supernatant was freeze-dried to obtain GnO/CN composite aerogel, followed by calcination at 550°C for 2 h in Ar to get rGnO/CN aerogel. The resultant aerogel delivered a marked electromagnetic shielding with a broad effective absorption bandwidth (EAB) of 8.08 GHz and a reflection loss (RL) of −38.08 dB. The electromagnetic parameters of the aerogels could be adjusted by controlling the weight ratio of *g*-CN to GnO, as well as the calcination temperature.

A heterojunction-structured photocatalyst, 3,4-dihydroxybenzaldehyde (DBD)-functionalized Ga_2O_3/*g*-CN, was synthesized by direct thermal polymerization of Ga_2O_3-DBD nanoparticles and urea at 550°C for 2 h and coded as Ga_2O_3-DBD_X/*g*-CN, where *X* corresponds to the weight percentage of Ga_2O_3 [87]. The preparation process is illustrated in Figure 2.27. In the composite, the aromatic rings were expected to contribute to the formation of a well-developed combined interface

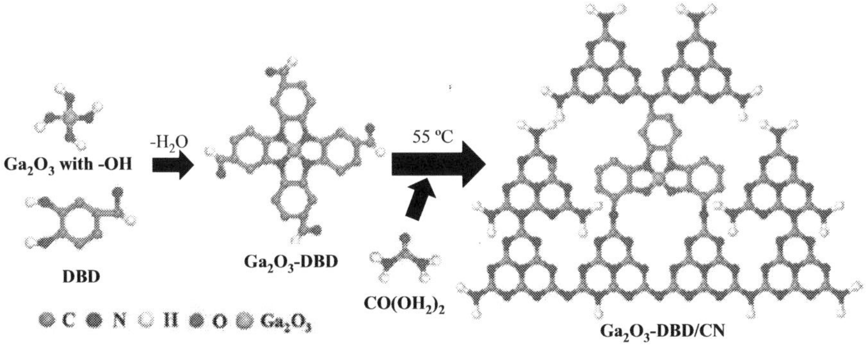

FIGURE 2.27 Procedure for the preparation of Ga_2O_3-DBD/CN composite [87].

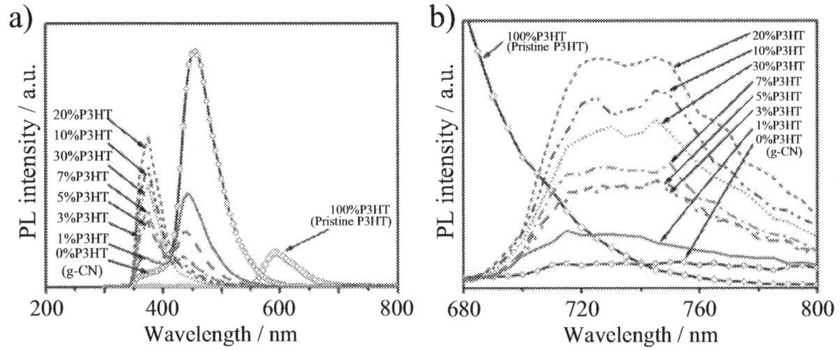

FIGURE 2.28 Photoluminescence spectra of P3HT-loaded CNs with different P3HT contents: (a) wavelength range of 200–800 nm and (b) wavelength range of 680–800 nm [88].

between Ga_2O_3-DBD and g-CN sheet. The composite exhibited a wide absorption range, high charge-separation efficiency, and high redox potential and, as a consequence, the enhancements in its activity and stability for visible-light photocatalytic N_2-fixation.

2.4.4 With Organics

A polymeric semiconductor, poly(3-hexylthiophene) (P3HT) (0.0–30.0 wt%), was loaded onto g-CN in different P3HT contents by immersing CN powder in a chloroform solution of P3HT overnight [88]. P3HT is known to be a polymer semiconductor with a small bandgap of 1.9–2.1 eV, a high charge carrier mobility, a dissolubility, a processability, and a long-term stability. The PL spectra of the composite catalysts under incident light with a wavelength of 259 nm are shown in Figure 2.28a and b in two wavelength ranges, 200–800 nm and 680–800 nm. The peaks at 453 and 593 nm can be attributed to electron–hole recombination of g-C_3N_4 and P3HT, respectively. Two new PL emission peaks at 375 nm and 720–750 nm are observed in the spectra of the composite catalysts, of which the intensity increases with an increase in the

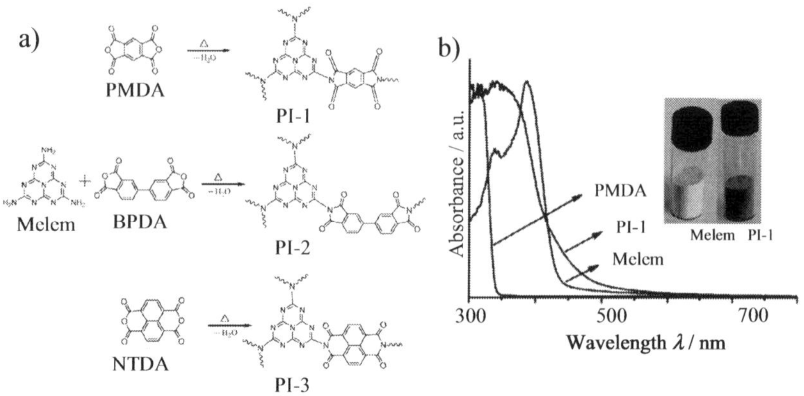

FIGURE 2.29 Composites of melem with three kinds of di-anhydrides PMDA, BPDA, and NTDA (PI-1, PI-2, and PI-3, respectively): (a) Reaction scheme for the synthesis of PIs and (b) UV-vis spectra of the PI-1 in comparison with the precursor melem and PMDA [52].

amount of P3HT and reaches a maximum when the amount of P3HT is approximately 20 wt%. The photon energy of 375 nm (3.3 eV) agrees with the value of 3.1 eV for the difference between the CB of P3HT and the VB of *g*-CN, while that of 740 nm corresponds to the difference between the CB of *g*-CN and the VB of P3HT.

The loading of 7,7,8,8-tetracyanoquinodimethane (TCNQ) was done onto *g*-CN by mixing a well-dispersed *g*-CN with the TCNQ-DMF solution (concentration of 5 mg/mL) in water under sonication [89]. TCNQ is well known to have a highly conjugated system and abundant π electrons.

Three kinds of PI were synthesized by conjugating the melem with three kinds of di-anhydride monomers, i.e., PMDA, BPDA, and NTDA, in equal molar ratio by heating in a semi-closed system at 350°C (coded as PI-1, PI-2, and PI-3) [52]. The synthesis scheme of PIs is shown in Figure 2.29a. As an example, the UV-vis spectrum of PI-1 in comparison with the precursors, melem and PMDA, is shown with the photographs of the melem and the resultant PI-1 in Figure 2.29b. The catalyst PI-1 exhibited a significant improvement in the photocatalytic H_2 evolution over the precursor melem under visible light irradiation, which was ascribed to its drastically red-shifted optical absorption.

NH_2-UiO-66 is one of the most studied MOFs (metal-organic frameworks) for various photocatalytic applications due to its visible-light response and high thermal and chemical stability in an aqueous solution (UiO/IK-CN composite) [90]. It matches in the band structures with I^- and K^+ co-doped CN (IK-CN) to form Z-scheme heterojunctions for the enhancement of H_2O_2 production [91].

g-CN particle-anchoring onto bacterial cellulose (BC) aerogels was successfully performed because of the sufficient surface hydroxyl groups and high porosity of BC aerogel [92]. The composite was prepared by mixing a melamine-derived *g*-CN powder (0.06, 0.12, and 0.18 g) in 60 g of 0.2wt% BC dispersion under ultrasonication, being a weight ratio *g*-CN/BC of 0.5/1, 1/1, and 1.5/1 and coded as CN/BC-1, -2, and -3, respectively. The *g*-CN content in the composites was calculated to be 34.1, 47.8,

TABLE 2.7
Crystal and Textural Properties, Elemental Composition, and Bandgaps of COF/CN Composites [93]

COF-content (wt%)	d_{002} (nm)	$Lc(002)$ (nm)	S_{BET} (m^2/g)	V_{total} (cm^3/g)	Elemental Composition C (wt%)	N (wt%)	S (wt%)	N/C	Bandgap (eV)
0	0.3248	12.9	7.7	0.034	34.2	59.7	0.0	1.50	2.55
5	0.3250	15.8	19.2	0.082	36.0	59.0	0.0	1.40	2.55
10	0.3245	11.9	23.4	0.101	38.5	56.4	0.2	1.26	2.50
20	0.3252	12.2	30.1	0.102	39.8	54.4	1.0	1.17	2.35
25	0.3238	14.7	16.4	0.069	37.1	54.5	1.9	1.26	2.35
100	–	–	36.2	0.0191	46.0	39.0	7.4	–	2.20

and 59.3 wt%, respectively. The composites were obtained as disks with a diameter and thickness of approximately 4.9 and 0.5 cm, of which the bulk density was kept to be 2.6–11.5 kg/m^3, and applied to an air filter.

A composite of the melamine-derived g-CN with a COF was synthesized using microwave-assisted solvothermal treatment of the mixture of g-CN with 5–25 wt% COF in DMF at 120°C for 30 min [93]. For comparison, the g-CN was also submitted to microwave treatment following the same steps without COF. The COF was prepared by refluxing a solution of mixed tri-(4-formylphenoxy) cyanurate (TFPC) with 1,5-diaminonaphthalene, sulfur, DMSO, and DMF at 140°C for 24 h. In the resultant composites with heterojunctions, the CN-component retained high crystallinity, showing a sharp 002 diffraction peak, while an unsymmetrical 10 peak. Crystal and textural properties measured by XRD and N_2 adsorption isotherm, elemental composition by chemical analysis, and bandgap calculated from UV-vis spectrum are summarized in Table 2.7. On these composites, the degradation of acetaminophen (an antipyretic) in water was studied.

2.5 C_3N_5-BASED MATERIALS

Carbon nitride framework with a C_3N_5 stoichiometry was synthesized by thermal de-ammoniation of 2,5,8-trihydrazino-s-heptazine (melem hydrazine, MH) at a temperature of 450°C for 2 h [94]. The precursor MH was prepared by the hydrothermal reaction between melem and hydrazine hydrate at 140°C for 24 h. In Figure 2.30, the synthesis procedure with atomic structures of melem, MH, and C_3N_5 is shown. C_3N_5 is composed of heptazine moieties bridged together by azo linkage (–N=N–) of which the presence was confirmed by ^{15}N NMR spectrum at −271 ppm. The C_3N_5 constitutes an entirely new and different bonding fashion from g-C_3N_4, where three heptazine units are linked together with tertiary nitrogen. The resultant C_3N_5 has the reduced bandgap of 1.76 eV. It could adsorb methylene blue (MB) larger amount than g-C_3N_4 does, while C_3N_5 could not degrade MB under simulated sunlight.

The square-hole (macropore) g-C_3N_5 (sh-C_3N_5) was synthesized using KBr-templating [95]. The mixture of KBr with 3-amino-1,2,4-triazole in a weight

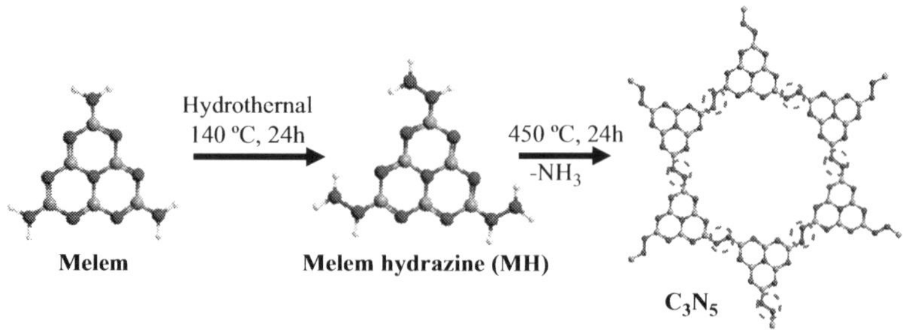

FIGURE 2.30 Synthesis process of C_3N_5 with atomic structures. On C_3N_5, the presence of azo linkages is highlighted [94].

ratio of 10/1.5 was prepared by dispersing in water under stirring. The mixture was dried at 60°C, calcined at 550°C for 3h in air and then washed with water until no Br^- could be detected to produce *sh*-C_3N_5. N_2-fixation performance of *sh*-C_3N_5 using 50mg of catalysts was compared with the composite of V^{4+}-doped carbon dots (V-CD). V-CD was synthesized from a mixture of chitosan oligosaccharide with 1,10-phenanthroline ($C_{12}H_8N_2$) and V^{4+}-oxyacetylacetonate ($C_{10}H_{14}O_5V$) via the solvothermal procedure at 160°C for 12h. The introduction of square-holes (square-shaped macropores) in *g*-C_3N_5 and its compositing with V-CD were experimentally confirmed to result in a marked enhancement in N_2-fixation performances.

2.6 CONCLUDING REMARKS

Various nitrogen- and carbon-containing compounds have been used as the precursor for the synthesis of *g*-CN, including urea, melamine, cyanamide, and DCDA, which were heat-treated in a ceramic crucible, such as an alumina crucible, with a lid at around 500–600°C. The texture of the bulk *g*-CN also influenced its photocatalytic activity and was modified by various techniques, including grinding, templating, and exfoliation.

To conclude this chapter, tuning techniques for *g*-CN-based materials are summarized here, which may present the advantages of *g*-CN-based materials and may characterize the *g*-CN materials. The techniques for tuning the photoactivity, in other words, the bandgap, of the *g*-CN-based materials, were classified here into three categories:

a. Doping of non-metallic atoms, metallic atoms, and small molecules.
b. Templating using silica, etc., to control pore structure.
c. Compositing (loading of inorganic and organic molecules).

2.6.1 Doping

Doping of various atoms was reviewed to be an efficient method to tune the unique electronic structure and bandgap of *g*-CN, which considerably broadens the light-responsive range and enhances the charge separation. The development of efficient, low-cost doped g-CN for use in various applications such as photocatalytic H_2 evolution, CO_2 conversion, and removal of pollutants in water and gas was summarized [96].

2.6.1.1 Nonmetal Atoms

The study based on the first principle calculation demonstrated that nonmetal doping, such as P, B, C, O, and S, was an effective way to enhance the photocatalytic performance of g-CNs [97]. Most nonmetal atoms can be introduced into the monolayer g-CN. The cell parameters of CN are altered due to the bond lengths or interaction strengths of the doped foreign atoms to C and N, associating with the bond relaxation around them, which affects the electronic structure and surely photocatalytic performances of the system. In Table 2.8, cell parameter a_0 and layer thickness h on g-CN with some nonmetal dopants are listed together with the bandgap E_g, revealing a marked increase in h, reaching even three times that of the pristine, while a_0 and bandgap E_g change only slightly. The electronic states of the monolayer g-CN after the doping of some nonmetal atoms (B, O, S, P, I, etc.) are calculated by comparing with those of pristine g-CN. O-doping into g-CN using formaldehyde resulted in the O-linked g-CN [98], and P-doping using PCT [51] resulted in the O-linking and P-linking of heptazines.

B-doping into g-CN was performed using two different B-precursors, BH_3NH_3 [68] and boron oxide [99]. In the former, the B-atom is supposed to replace the N-atom connecting the triazine unit, as predicted by DFT calculation and shown in Figure 2.20c [67], while the latter B-atom is doped as a functional group at the edge of the triazine framework. In the former case, B-atom-replacing for N-atom in the framework strongly distorts the layer stacking, as a consequence, the 002 peak becomes broader with the increase in B-content from 0.5 to 3.0 at%, and even the two-dimensional 10 peak becomes non-detectable. In the latter case, however, B-doping distortion gives an influence on layer stacking, as a consequence, no detectable changes in XRD profiles of 002 and 10 peaks. In both cases, the photocatalytic activity of the CN-based materials was enhanced by B-doping.

TABLE 2.8
Cell Parameters *a* and Layer Thickness *h* of g-CNs Doped by Some Nonmetals [97]

g-CN	a_0 (Å) (Ratio to the Pristine)	h (Å) (Ratio to the Pristine)	Bandgap E_g (eV)
Pristine	6.960	1.379	1.89
B-doped	7.093 (1.02)	2.682 (1.94)	1.86
O-doped	6.900 (0.98)	2.240 (1.62)	1.73
S-doped	7.180 (1.03)	2.781 (2.02)	1.88
P-doped	7.174 (1.03)	2.257 (2.64)	1.86
I-doped	6.820 (0.98)	3.206 (3.32)	1.83

The effects of the doping of nonmetals, such as B, P, and sulfuric acid (SA, $-SO_3H$), into *g*-CN on the structure and physicochemical properties together with photocatalytic activity for the oxidative desulfurization, were studied [100]. The results are summarized in Table 2.9 on the basis of the data reported in reference [100]. Selective oxidation of dibenzothiophene (DBT) with molecular O_2 as an oxidant under visible light irradiation was chosen for the photocatalytic activity of *g*-CN-based photocatalysts, using 20 mg photocatalysts dissolved in a 100 ppm DBT acetonitrile solution under blue LED (9 W) light under O_2 at 24 h. The optimization on the amount of the photocatalyst, the wavelength of visible light, and the irradiation time for the 100 ppm DBT acetonitrile solution of 100 mL was achieved on the basis of the experimental results on SA-CN to be 20 mg, 450–495 nm, and 24 h. The use of more than 20 mg of catalyst resulted in decreased DBT removal efficiency, mainly due to agglomeration of the catalyst particles.

The intercalation of NO_3^- into bulk *g*-CN (*b*-CN) and following exfoliation through heating at 450°C, 500°C, 530°C, and 570°C was an effective method to obtain ultrathin sheets of *g*-CN (*fl*-CN-*X*, where *X* corresponds to exfoliation temperature) [63]. Ultrathin *g*-CN sheet obtained at 530°C (*fl*-CN-530) had an eight-layered structure. The *fl*-CN catalysts also exhibited enhanced activity for H_2O_2 production in photocatalytic O_2 reduction, achieving a high H_2O_2-production rate, 952 μmol/g·h for *fl*-CN-530, 8.8 times that of the pristine *b*-CN. They also delivered a high activity for photocatalytic oxidation of CH_4 to form C1 oxygenates with a high selectivity of 99.2%. The catalyst *fl*-CN-530 gave a high production rate as 1,893 μmol/g·h under the reaction conditions: 10 mg of the catalyst, 20 mL of distilled water, 19 bar of CH_4, 1 bar of O_2, 25°C, under the irradiation of $300 < \lambda < 780$ nm, light intensity of ca. 200 mW/cm^2, stirring rate of 800 rpm, and reaction time of 3 h.

2.6.1.2 Metal Atoms

K-doped *g*-CN was synthesized from a dispersion of dicyandiamide ($C_2H_4N_4$) in KOH solution with 0.02, 0.05, 0.07, and 0.09 M concentration and annealing at 520°C for 2 h after the removal of water at 100°C and milling [101]. The resulting K-doped CN powders were coded as K_X-CNs, where *X* corresponds to the KOH concentration used. A possible site for K^+ in *g*-CN framework is the intrinsic vacancy V_{6N} and, as a consequence, no changes in the XRD pattern and N_2 adsorption–desorption isotherm at 77 K were detected. However, since K^+ is accommodated at the center of V_{6N} with the interaction between positive K^+ and lone pair electrons of N, like chelating, the electronic band structure and bandgap depend strongly on the K^+-dosage, as shown in Figure 2.31, suggesting the strong dependence of photocatalytic activities.

Na and K co-doped *g*-CN was synthesized from melamine using KCl-NaCl eutectic mixture (1/1 by mole, melting temperature of 476°C) as solvent at 520°C for 2 h [42]. Increasing the relative ratio of the eutectic mixture used (i.e., increasing the dosages of Na and K) had no significant effect on the structural parameters (i.e., the broadening of the 002 peak and the difficulty in detecting the 10 peak), while the PL spectrum was significantly weakened and exhibited increased photocatalytic activity. In contrast, the synthesis of *g*-CN using LiCl-KCl eutectics was reported to result in highly crystalline *g*-CN by showing a full-range powder XRD profile and high-resolution TEM images [41]. Neither Li- nor K-doping was detected.

TABLE 2.9
Structure, Physicochemical Properties of the Pristine, B-Doped, P-Doped, and SA-Doped *g*-CNs Together with DBT-Removability. Produced on the Basis of the Data Reported in Reference [100]

Catalyst	Pristine *g*-CN (Pristine)	B-Doped *g*-CN (B-CN)	P-Doped *g*-CN (P-CN)	SA-Doped *g*-CN (SA-CN)
CN precursor	Melamine	Melamine	Melamine	Pristine *g*-CN
Synthesis process	In an Al_2O_3 crucible at 520°C for 2h	Mixed with BH_3O_3 At 550°C for 2h	Mixed with $NH_4H_2PO_2$ At 570°C for 3h	Dissolved into H_2SO_4 solution Refluxing at 135°C for 24h
Structure	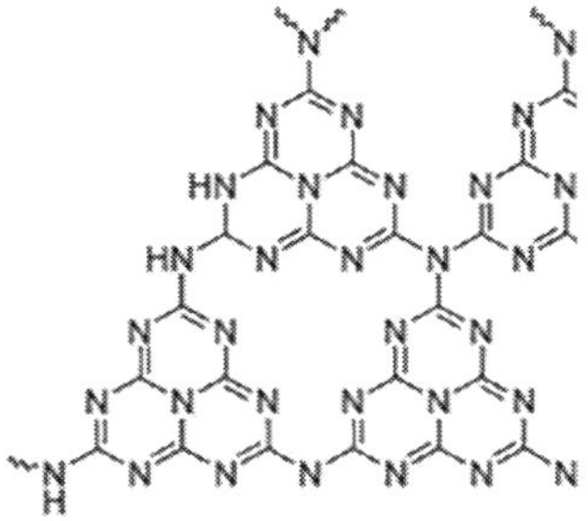	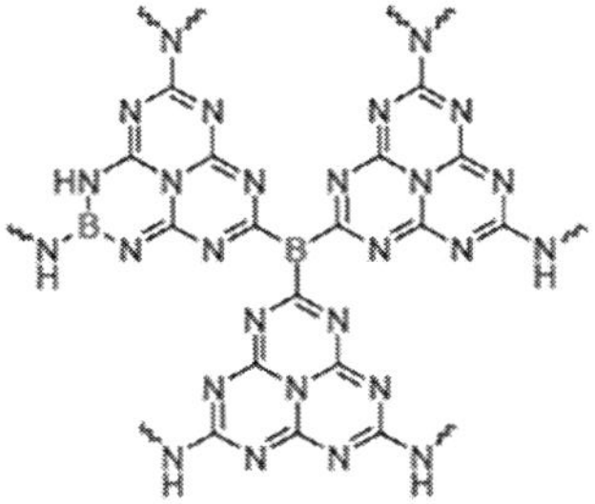	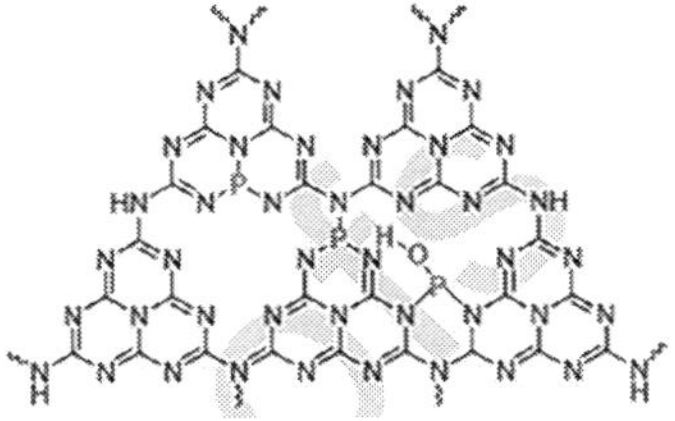	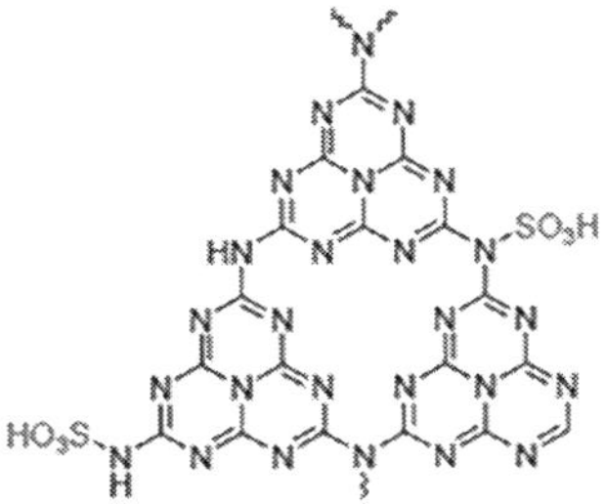
XRD profile	Sharp 002 peak with broad 100 peak	Slightly broad 002 peak	Broad 002 peak	Broad 002 peak
Pore structure	S_{BET}: 8.16 m^2/g V_{total}: 0.053 cm^3/g	S_{BET}: 3.53 m^2/g V_{total}: 0.020 cm^3/g	S_{BET}: 69.55 m^2/g V_{total}: 0.1819 cm^3/g	S_{BET}: 16.22 m^2/g V_{total}: 0.083 cm^3/g
Bandgap	2.76eV	2.67eV	2.72eV	2.80eV
Removal of DBT	76%	20%	15%	93%

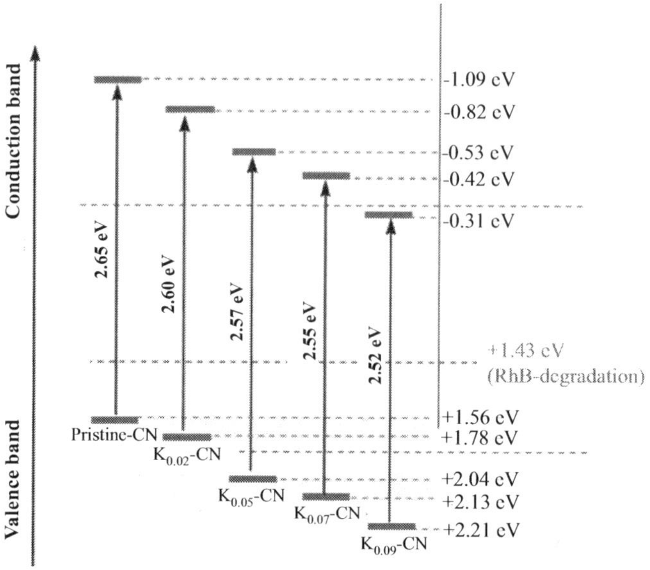

FIGURE 2.31 The band structures of K-doped *g*-CNs (K_X-CN) and bandgaps referring to the potential level for RhB degradation [101].

2.6.2 Templating

A hard template of silica SiO_2 was used to prepare *mpg*-CNs from two mixtures of ethylenediamine and carbon tetrachloride as CN-precursors with ordered mesoporous silica (SBA-15) [102], of cyanamide with SBA-15 [103,104], and also of cyanamide with tetraethyl-orthosilicate (TEOS, SiO_2-precursor) [60]. The N_2 adsorption–desorption isotherms and pore-size distributions calculated from these isotherms demonstrated the formation of mesopores in the resulting *g*-CNs, while the XRD patterns are comparable with those of the pristine *g*-CN.

A simultaneous application of a soft template (surfactant F127) and a hard template (silica sol) was carried out for creating the pores in *g*-CN [105]. An aqueous solution of melamine was added to silica sol (7 nm size) suspension (30 wt%) and then different amounts of a commercial surfactant F127 (0, 0.05, 0.25, 1.00, 1.25, and 1.5 g) were added under strong stirring, followed by drying at 80°C for 12 h and calcining at 550°C for 4 h. The resultant solids were coded as F_XM/CN, where X (0–6) corresponds to the amount of F127 added. For comparison, the pristine CN was prepared without SiO_2 and F127, M/CN using only silica-sol template (5 mL of 30 wt% SiO_2) and F/CN using only F127 (1 g). The pore structure parameters obtained from N_2-adsorption isotherms are listed in Table 2.10 revealing that the addition of F127 more than 1 g (F_4M/CN) resulted in the formation of large pores. The introduction of large pores in the *g*-CN catalyst resulted in a significant acceleration of the methyl orange degradation reaction under a UV light from a 15 W lamp with a wavelength of 405 nm. The catalyst F_4M/CN delivered much faster degradation than the *mpg*-CN (i.e., M/CN) and the addition of the soft template F127 (F/CN) to a hard template (silica sol) demonstrated a cooperative effect of two templates.

TABLE 2.10
Pore Structure Parameters of the Resultant *g*-CNs [105]

g-CNs	F127-Addition (g)[a]	SiO_2-Addition (mL)[b]	S_{BET} (m^2/g)	V_{total} (cm^3/g)	Average Pore Size D_a (nm)	Maximum Pore Size D_{max} (nm)
Pristine	–	–	5.00	0.025	19.87	–
M/CN	–	5	13.32	0.046	13.92	–
F/CN	1.0	–	10.87	0.060	21.91	–
F_1MCN	0.05	5	9.14	0.037	15.17	–
F_2MCN	0.25	5	12.38	0.047	9.4	4.10
F_3MCN	0.5	5	81.99	0.192	16.10	4.00/7.88
F_4MCN	1.0	5	104.18	0.346	13.27	11.86/16.13
F_5MCN	1.25	5	97.74	0.401	16.41	25.99
F_6MCN	1.5	5	86.10	0.311	14.43	15.24/24.40

[a] Amount of F127 for 5 g *g*-CN.
[b] Amount of silica (30 wt% sol).

2.6.3 Compositing

It must be pointed out that *g*-CN itself has photocatalytic activity and also works as an excellent scaffold for other photocatalysts. The loading of other catalysts enhances not only *g*-CN's own activity but also enhances the photoactivity of the loaded catalysts. In addition, *g*-CN could support single metal atoms to obtain enhanced catalytic activity of the metals and sufficient stabilities in cyclic uses of the catalysts [106].

Poly(3-hexylthiophene) (P3HT) is a polymer semiconductor with a bandgap of 1.9–2.1 eV, which was loaded onto *g*-CN via an impregnation of the melamine-derived *g*-CN into a chloroform solution of P3HT (0.6 mg/mL) overnight, followed by evaporation of the solvent to get composite P3HT/CN [88]. The H_2-evolution performances were investigated by dispersing 0.3 g of catalyst after 1 wt% Pt deposition in a water containing 0.25 M Na_2S and 0.25 M Na_2SO_3 (600 mL). The resultant composite P3HT/CN delivered a markedly enhanced H_2-evolution rate, reaching approximately 300 times faster than that of the pristine CN and P3HT. The activity of the composite depends strongly on the P3HT content in the composite and the optimum content of P3HT was shown to be 3 wt%. A high-resolution TEM image of the 3 wt%P3HT/CN composite showed the lattice fringe spacing of ca. 1.64 nm corresponding to the (100) plane of the crystalline P3HT.

The triazine-based COF coded as TFPT/DHTH/COF was synthesized by condensation of precursors 1,3,5-tris-(4-formyl-phenyl) triazine (TFPT) and 2,5-bis-(3-hydroxypropoxy) terephthalohydrazide (DHTH), on the surface of which CdS nanoparticles were deposited [82]. The resultant composite exhibited a high H_2-evolution rate (15.75 mmol/h·g), which is approximately 75 times that of the component of the composite, i.e., TFPTDHTH-COF without CdS-deposition (0.21 mmol/h·g) and 3.4 times that of CdS (4.57 mmol/hg).

TABLE 2.11
Photocatalytic H_2-Evolution Performances with S_{BET} and Bandgap for PI Catalysts [52]

Photocatalyst	Di-Anhydride	S_{BET} (m^2/g)	Bandgap (eV)	H_2-Evolution Rate (µmol/h)		Specific H_2-Evolution Rate ($mmol/m^2 \cdot h$)	
				$\lambda > 300$ nm	$\lambda > 420$ nm	$\lambda > 300$ nm	$\lambda > 420$ nm
Melem	–	4.3	2.9	64.4	1.9	15.0	0.5
PI-1	PMDA	8.6	2.5	55.0	13.1	6.4	1.5
PI-2	BPDA	1.1	2.8	22.0	1.2	20.0	1.1
PI-3	NTDA	5.2	1.9	10.9	1.3	2.1	0.3
g-CN	–	10	2.7	93.8	7.0	9.4	0.7

Three kinds of polyimides PI-1, -2, and -3 were synthesized by conjugating melem with three kinds of di-anhydride monomers, PMDA, BPDA, and NTDA [32]. The H_2-evolution rates over the catalysts are listed together with S_{BET} and the bandgap in Table 2.11. PI-2 and PI-3 exhibit a lower H_2-evolution activity than the pristine melem under the irradiation of either full arc (>300 nm) or visible light (>420 nm), while PI-1 exhibits higher activity than the pristine melem under visible ligh irradiation condition.

Cu atoms are uniformly anchored on the surface of sulfur-doped *g*-CN (S-CN) via a phytate acid (PA)-bridging (CuPA/S-CN) for achieving an enhanced CO_2 conversion to CH_3OH [58]. The product from the CO_2-conversion reaction was mainly CH_3OH, and its rate was markedly enhanced and it was shown that 15% loading of CuPA was the optimum.

REFERENCES

1. Ong W-J, Tan L-L, Ng YH, et al., Graphitic carbon nitride (*g*-C_3N_4)-based photocatalysts for artificial photosynthesis and environmental remediation: Are we a step closer to achieving sustainability? *Chem Rev* 2016; 116: 7159–329.
2. Kang Y, Yang Y, Yin L-C, et al., An amorphous carbon nitride photocatalyst with greatly extended visible-light-responsive range for photocatalytic hydrogen generation. *Adv Mater* 2015; 27: 4572–7.
3. Zheng Y, Luo Y, Ruan Q, et al., Plasma-induced hierarchical amorphous carbon nitride nanostructure with two N_2C-site vacancies for photocatalytic H_2O_2 production. *Appl Catal B* 2022; 311: 121372.
4. Liu Y, Chen X, Kamali M, et al., Unraveling the presence and positions of nitrogen defects in defective *g*-C_3N_4 for improved organic photocatalytic degradation: Insights from experiments and theoretical calculations. *Adv Funct Mater* 2024; 34: 2405741.
5. Li L, Liu H, Cheng C, et al., Photochemical tuning of tri-coordinated nitrogen deficiency in carbon nitride to create delocalized π electron clouds for efficient CO_2 photoreduction. *ACS Catal* 2024; 14: 10204–13.
6. Nakagawa T, Fujiwara S, Shimoda S, et al., Photocatalytic formaldehyde decomposition efficiency of *g*-C_3N_4 prepared from melamine/urea mixtures: Influence of starting material composition. *Chem Phys Lett* 2024; 850: 141457.

7. Negro P, Cesano F, Damin A, et al., Porous g-C_3N_4-based nanoarchitectures by playing with sustainable precursors: Role of urea/melamine ratio on the structure/properties relationship. *J Alloys Compounds* 2024; 1002: 175053.
8. Lv C, Qian Y, Yan C, et al., Defect engineering metal-free polymeric carbon nitride electrocatalyst for effective nitrogen fixation under ambient conditions. *Angew Chem* 2018, 130, 10403–7.
9. Mondal S, Mark G, Tashakory A, et al., Porous carbon nitride rods as an efficient photoanode for water splitting and benzylamine oxidation. *J Mater Chem A* 2024; 12: 11502–10.
10. Gong Y, Yang P, Ma D, et al., Construction of three-coordinated (N_3C) nitrogen vacancies in g-C_3N_4 for efficient photocatalytic CO_2 reduction. *Ceram Int* 2024; 50(18B): 33131–42.
11. Rao F, Zhong J, Li J, Improved visible light responsive photocatalytic hydrogen production over g-C_3N_4 with rich carbon vacancies. *Ceram Int* 2022; 48: 1439–45.
12. Shao S, Liu X, Wang R, et al., Morphologic and microstructural modulation of graphitic carbon nitride through EDTA-2Na mediated supramolecular self-assembly route: Enhanced visible-light-driven photocatalytic activity for antibiotic degradation. *Appl Surf Sci* 2024; 669: 160501.
13. Jun Y-S, Park J, Lee SU, et al, Three-dimensional macroscopic assemblies of low-dimensional carbon nitrides for enhanced hydrogen evolution. *Angew Chem Int Ed* 2013; 52: 11083–7.
14. Bojdys MJ, Muller J, Antonietti M, et al., Ionothermal synthesis of crystalline, condensed, graphitic carbon nitride. *Chem Eur J* 2008; 14: 8177–82.
15. Liu J, Fu W, Liao Y, et al., Recent advances in crystalline carbon nitride for photocatalysis. *J Mater Sci Technol* 2021; 91: 224–40.
16. Li Z, Ma L, Yu M, et al., Crystalline graphitic carbon nitride in photocatalysis. *Surf Interfaces* 2024; 51: 104492.
17. Hatta MHM, Lintang HO, Lee SL, et al., Synthesis of highly active crystalline carbon nitride prepared in various salt melts for photocatalytic degradation of phenol. *Turk J Chem* 2019; 43: 63–72.
18. Xu Y, He X, Zhong H, et al., Solid salt confinement effect: An effective strategy to fabricate high crystalline polymer carbon nitride for enhanced photocatalytic hydrogen evolution. *Appl Catal B* 2019;246: 349–55.
19. Zhang G, Xu Y, Zhu J, et al., Enhanced photocatalytic H_2 production independent of exciton dissociation in crystalline carbon nitride, *Appl Catal B* 2023; 338: 123049.
20. Xu T, Hur J, Niu P, et al., Synthesis of crystalline g-C_3N_4 with rock/molten salts for efficient photocatalysis and piezocatalysis, *Green Energy Environ* 2024; 9: 890–8.
21. Zhuge G-Y, Zhang W-D, Boosting photocatalytic hydrogen evolution rate over carbon nitride through tuning its crystallinity and its nitrogen composition. *J Colloid Interface Sci* 2019; 555: 268–75.
22. Liang Z, Zhuang X, Tang Z, et al., Soft-template induced synthesis of high-crystalline polymeric carbon nitride with boosted photocatalytic performance, *J Mater Chem A* 2021; 9: 6805–10.
23. Xing W, Tu W, Han Z, et al., Template-induced high crystalline g-C_3N_4 nanosheets for enhanced photocatalytic H_2 evolution, *ACS Energy Lett* 2018; 3: 514–9.
24. Wu X, Ma H, Zhong W, et al., Porous crystalline g-C_3N_4: Bifunctional $NaHCO_3$ template-mediated synthesis and improved photocatalytic H_2-s. *App Catal B* 2020; 271: 118899.
25. Wu X, Cheng J, Li X, et al., Enhanced visible photocatalytic oxidation of NO by repeated calcination of g-C_3N_4. *Appl Surf Sci* 2019; 465: 1037–46.
26. Cheng J, Hu Z, Lv K, et al., Drastic promoting the visible photoreactivity of layered carbon nitride by polymerization of dicyandiamide at high pressure. *Appl Catal B* 2018; 232: 330–9.

27. Lin X, Du H, Jiang D, et al., Microwave awakening the n–π* electronic transition in highly crystalline polymeric carbon nitride nanosheets for photocatalytic hydrogen generation. *J Energy Chem* 2022; 65: 541–7.
28. Mazzanti S, Savateev A, Emerging concepts in carbon nitride organic photocatalysis. *ChemPlusChem* 2020; 85: 2499–517.
29. Wirnhier E, Döblinger M, Gunzelmann D, et al., Poly(triazine imide) with intercalation of lithium and chloride ions $(C_3N_3)_2(NH_xLi_{1-x})_3{\cdot}LiCl]$: A crystalline 2D carbon nitride network. *Chem A Eur J* 2011; 17: 3213–21.
30. Villalobos LF, Vahdat MT, Dakhchoune M, et al., Large-scale synthesis of crystalline g-C_3N_4 nanosheets and high-temperature H_2 sieving from assembled films. *Sci Adv* 2020; 6(4): 9851.
31. Xu W, An X, Zhang Q, et al., Cesium salts as mild chemical scissors to trim carbon nitride for photocatalytic H_2 evolution, *ACS Sustain Chem Eng* 2019; 7: 12351–7.
32. Song H, Luo L, Wang S, et al., Advances in poly(heptazine imide)/poly(triazine imide) photocatalyst. *Chin Chem Lett* 2024; 35(10): 109347.
33. Guo Y, Li J, Yuan Y, et al., A rapid microwave-assisted thermolysis route to highly crystalline carbon nitrides for efficient hydrogen generation. *Angew Chem Int Ed* 2016; 55: 14693–7.
34. Shen C, Lou Q, Liu K, et al., Microwave-assisted confining growth and liquid exfoliation of sp^3-hybrid carbon nitride nano/micro-crystals. *Energy Environ Mater* 2024; 7: e12772.
35. Du Y, Nie K, Wang W, et al., Bandgap modulations and promoted carrier separation in crystalline g-C_3N_4 thin films via chemical vapor deposition. *Appl Phys Lett* 2024; 124: 222101.
36. Chen L, Yan R, Oschatz M, et al., Ultrathin 2D graphitic carbon nitride on metal films: Underpotential sodium deposition in adlayers for sodium-ion batteries. *Angew Chem Int Ed* 2020; 59: 9067.
37. Giusto P, Cruz D, Heil T, et al., Shine bright like a diamond: New light on an old polymeric semiconductor. *Adv Mater* 2020; 32: 1908140.
38. Shen D, Imbault AL, Balati G, et al., Dissolution of g-C_3N_4 using zinc chloride molten salt hydrates for nanobelt fabrication and photocatalytic H_2O_2 production. *Chem Eur J* 2024; 30(48): e202401847.
39. Zhang Z, Leinenweber K, Bauer M, et al., High pressure bulk synthesis of crystalline $C_6N_9H_3{\cdot}HCl$: A novel C_3N_4 graphitic derivative. *J Am Chem Soc* 2001; 123: 7788–96.
40. Lin L, Lin Z, Zhang J, et al., Molecular-level insights on the reactive facet of carbon nitride single crystals photocatalysing overall water splitting. *Nat Catal* 2020; 3: 649–55.
41. Chong SY, Jones James TA, Khimyak YZ, et al., Tuning of gallery heights in a crystalline 2D carbon nitride network. *J Mater Chem A* 2013; 1: 1102–7
42. Zhao J, Ma L, Wang H, et al., Novel band gap-tunable K-Na co-doped graphitic carbon nitride prepared by molten salt method. *Appl Surf Sci* 2015; 332: 625–30.
43. Schwinghammer K, Mesch MB, Duppel V, et al., Crystalline carbon nitride nanosheets for improved visible-light hydrogen evolution. *J Am Chem Soc* 2014; 136: 1730–3.
44. Wang Y, Zhang H, Wei D, et al., Molten-salt-induced structural heterogeneity of carbon nitride for efficient photocatalytic H_2O_2 production. *Chem Commun* 2024; 60: 10732.
45. Yang D, Jia R, Wang X, et al., Highly ordered C-rich carbon nitride nanomaterials with pyrrole and imidazole moieties for photocatalytic overall water splitting through built-in electric field. *ACS Appl Nano Mater* 2024; 7(11): 12620–30.
46. Kim IY, Kim S, Jin X, et al., Ordered mesoporous C_3N_5 with a combined triazole and triazine framework and its graphene hybrids for the oxygen reduction reaction (ORR). *Angew Chem Int Ed* 2018; 57(52): 17135–40.

47. Zhang J, Chen X, Takanabe K, et al., Synthesis of a carbon nitride structure for visible-light catalysis by copolymerization. *Angew Chem Int Ed* 2010; 49: 441–4.
48. Zhang Q, Zhou J, Zhang H, et al., Terphenyl functionalized covalent triazine polymer as metal-free photocatalyst for superior hydrogen peroxide production. *Adv Funct Mater* 2024; 32(32): 2401579.
49. Ranganathan A, Pedireddi VR, Sanjayan G, et al., Sensitive dependence of the hydrogen-bonded assemblies in cyanuric acid – 4,4′-bipyridyl adducts on the solvent and the structure of the parent acid. *J Mol Struct* 2000; 522: 87–94.
50. Schwinghammer K, Tuffy B, Mesch MB, et al., Triazine-based carbon nitrides for visible-light- driven hydrogen evolution. *Angew Chem Int Ed* 2013; 52: 2435–9.
51. Song M, Song H, Chen P, et al., Polarization reorientation of carbon nitride induced by a local bridge donor for boosting photocatalytic H_2 evolution integrated with selective amine oxidation. *ACS Sustain Chem Eng* 2023; 11: 8096–195.
52. Chu S, Wang C, Feng J, et al., Melem: A metal-free unit for photocatalytic hydrogen volution. *Int J Hydrogen Energy* 2014; 39: 13519–26.
53. Simanek EE, Mammen M, Gordon DM, et al., Design and synthesis of hydrogen-bonded aggregates, theory and computation applied to three systems based on the cyanuric acid- melamine lattice. *Tetrahedron* 1995; 51: 607–19.
54. Shalom M, Inal S, Fettkenhauer C, et al., Improving carbon nitride photocatalysis by supramolecular preorganization of monomers. *J Am Chem Soc* 2013; 135: 7118–2.
55. Inagaki M, Itoi H, Kang F, *Porous Carbons: Syntheses and Applications.* Elsevier and Tsinghua University Press, Amsterdam, Netherlands and Beijing, China, 2022.
56. Kyotani T, Control of pore structure in carbon. *Carbon* 2000; 38: 269–86.
57. Inagaki M, Orikasa H, Morishita T, Morphology and pore control in carbon materials via templating. *RSC Adv* 2011; 1: 1620–40.
58. Goettmann F, Fischer A, Antonietti M, et al., Chemical synthesis of mesoporous carbon nitrides using hard templates and their use as a metal-free catalyst for Friedel-Crafts reaction of benzene. *Angew Chem Int Ed* 2006; 45: 4467–71.
59. Wang X, Maeda K, Chen XF, et al., Polymer semiconductors for artificial photosynthesis: Hydrogen evolution by mesoporous graphitic carbon nitride with visible light. *J Am Chem Soc* 2009; 131: 1680–1.
60. Kailasam K, Epping JD, Thomas A, et al., Mesoporous carbon nitride-silica composites by a combined sol-gel/thermal condensation approach and their application as photocatalysts. *Energy Environ Sci* 2011; 4: 4668–74.
61. Toyoda M, Hou· S, Huang Z- H, et al., Exfoliated graphite: Room temperature exfoliation and their applications. *Carbon Lett* 2022; 33: 335–62.
62. Huang J, Klahn M, Tian X, et al., Exfoliated polymeric carbon nitride nanosheets for photocatalytic applications. *ACS Appl Nano Mater* 2024; 7(7): 7442–52.
63. Feng B, Liu Y, Wan K, et al., Tailored exfoliation of polymeric carbon nitride for photocatalytic H_2O_2 production and CH_4 valorization mediated by O_2 activation. *Angew Chem Inter Ed* 2024; 63(18): e202401884.
64. Ma L, Guan R, Kang W, et al., Preparation of highly dispersed Ni single-atom doped ultrathin g-C_3N_4 nanosheets by metal vapor exfoliation for efficient photocatalytic CO_2 reduction. *J Colloid Interface Sci* 2024; 660: 381–92.
65. Zhang X, Liu Y, Ren M, et al., Precise carbon doping regulation of porous graphitic carbon nitride nanosheets enables elevated photocatalytic oxidation performance towards emerging organic pollutants. *Chem Eng J* 2022; 433: 134551.
66. Zhang S, An C, Zhang R, et al., Template-free synthesis of boron-doped graphitic carbon nitride porous nanotubes for enhanced photocatalytic hydrogen evolution. *Langmuir* 2024; 40(12): 6453–62.

67. Weng J, Zhang IY, Xu X, Identification of active sites on boron-doped graphitic carbon nitride as a metal-free single-atom photoelectrocatalyst for the nitrogen reduction reaction. *J Phys Chem C* 2024; 128(33): 13799–80.
68. Sagara N, Kamimura S, Tsubota T, et al., Photoelectrochemical CO_2 reduction by a P-type boron-doped g-C_3N_4 electrode under visible light. *Appl Catal B* 2016; 192: 193–8.
69. Ou H, Ning S, Zhu P, et al., Carbon nitride photocatalysts with integrated oxidation and reduction atomic active centers for improved CO_2 conversion. *Angew Chem Int Ed* 2022; 61: e202206579.
70. Kawahara T, Konishi Y, Tada H, et al., A patterned TiO_2 (Anatase)/TiO_2 (Rutile) bilayer-type photocatalyst: Effect of the Anatase/Rutile junction on the photocatalytic activity. *Angew Chem Int Ed* 2002; 41(15): 2811–3.
71. Khan MA, Rameel MI, Salam F, et al., Construction of isotype heterojunctions in polymeric carbon nitride with thermal modulation and improved photocatalytic hydrogen production activity. *Mater Chem Phys* 2024; 315: 129052.
72. Song H, Yang Y, Chen P, et al., Dual regulation platform in C_3N_4/$C_3N_{5.4}$ molecular heterojunctions for simultaneous photocatalytic oxidative and adsorption of arsenic. *Sep Purif Technol* 2024; 335: 126200.
73. Li H, Liu Y, Gao X, et al., Facile synthesis and enhanced visible-light photocatalysis of graphitic carbon nitride composite semiconductors. *ChemSusChem* 2015; 8: 1189–96.
74. Zhou Q, Guo Y, Ye Z, et al., Carbon nitride photocatalyst with internal electric field induced photogenerated carriers' spatial enrichment for enhanced photocatalytic water splitting. *Mater Today* 2022; 58: 100–9.
75. Balakrishnan A, Chinthala M, et al., Construction of dual Z-scheme ternary carbon nitride homojunction pectin microspheres as a multifunctional photocatalyst for tetracycline degradation, H_2O_2 production, and N_2 fixation. *Chem Eng J* 2024; 496: 153899.
76. Ma J, Peng X, Zhou Z, et al., Extended conjugation tuning carbon nitride for non-sacrificial H_2O_2 photosynthesis and hypoxic tumor therapy. *Angew Chem Int Ed* 2022; 61(43): e202210856.
77. Maeda K, Wang XC, Nishihara Y, et al., Photocatalytic activities of graphitic carbon nitride powder for water reduction and oxidation under visible light. *J Phys Chem C* 2009; 113: 4940–7.
78. Zhang Q, Wang B, Miao H, et al., Efficient photocatalytic H_2O_2 production over K^+-intercalated crystalline g-C_3N_4 with regulated oxygen reduction pathway. *Chem Eng J* 2024; 482: 148844.
79. Du W, Chen R, Yang L, et al., Phytic acid-bridged copper on sulfur-containing carbon nitride for enhancing photocatalytic CO_2 reduction to CH_3OH. *Ind Eng Chem Res* 2024; 63: 6201–9.
80. Li Q, Yue B, Iwai H, et al., Carbon nitride polymers sensitized with N-doped tantalic acid for visible light-induced photocatalytic hydrogen evolution. *J Phys Chem C* 2010; 114: 4100–5.
81. Tao FF, Dong Y, Yang L, NiO/g-C_3N_4 p–n heterojunctions wrapped by rGO for the enhanced CO_2 photocatalytic reduction. *ACS Sustain Chem Eng* 2024; 12(17): 6709–18
82. Zhang X, Wu X, Chen R, et al., A triazine-based covalent organic framework decorated with cadmium sulfide for efficient photocatalytic hydrogen evolution from water. *J Colloid Interface Sci* 2024; 665: 100–8.
83. Zhang K, Yuan H, Xu F, et al., Rational fabrication of MoS_2/g-C_3N_4 heterostructures for efficient photocatalytic degradation of Rhodamine B. *Chem Select* 2024; 9(30): e202401034.
84. Lin Y, Chen Z, Feng C, et al., Preparation of S-C_3N_4/AgCdS Z-scheme heterojunction photocatalyst and its effectively improved photocatalytic performance. *Molecules* 2024; 29(9): 1931.

85. Wang P, Wei S, Han Y, Lin S, et al., Metal-free C_{60}-doped mesoporous carbon nitride drives red-light photocatalysis. *Langmuir* 2024; 40(27): 14045–56.
86. Chen Z, Cao K, Zhang Y, et al., Self-assembled heterojunction interfaces between g-C_3N_4 and rGO nanosheets for enhanced electromagnetic wave absorption. *ACS Appl Nano Mater* 2024; 7(16):19688–97.
87. Cao S, Zhou N, Gao F, et al., All-solid-state Z-scheme 3,4-dihydroxybenzaldehyde functionalized Ga_2O_3/graphitic carbon nitride photocatalyst with aromatic rings as electron mediators for visible-light photocatalytic nitrogen fixation. *Appl Catal B* 2017; 218: 600–10.
88. Yan H, Huang Y, Polymer composites of carbon nitride and poly(3-hexylthiophene) to achieve enhanced hydrogen production from water under visible light. *Chem Commun* 2011; 47: 4168–70.
89. Zhang M, Yao WQ, Lv YH, et al., Enhancement of mineralization ability of C_3N_4 via a lower valence position by a tetracyanoquinodimethane organic semiconductor. *J Mater Chem A* 2014; 2: 11432–8.
90. Lei Y, Zhu L, Xu J, Liu S, et al., The metal organic framework of UiO-66-NH_2 reinforced nanofiltration membrane for highly efficient ion sieving. *J Environ Chem Eng* 2023; 11(6): 111222.
91. Lu X, Chen Z, Hu Z, et al., Boosted charge transfer for highly efficient photosynthesis of H_2O_2 over Z-scheme I^-/K^+ co-doped g-C_3N_4/metal-organic-frameworks in pure water under visible light. *Adv Energy Mater* 2024; 14(38): 2401873
92. Li M, Zhang XF, Qiu J, et al., Flexible g-C_3N_4@bacterial cellulose aerogels for integrated indoor air purification. *Colloids Surf A* 2024; 698: 134567.
93. Zandipak R, Bahramifar N, Torabi M, et al., Synergistic effect of graphitic-like carbon nitride and sulfur-based thiazole-linked organic polymer heterostructures for boosting the photocatalytic degradation of pharmaceuticals in water. *Chem Eng J* 2024; 494: 152843.
94. Kumar P, Vahidzadeh E, Thakur UK, et al., C_3N_5: A low bandgap semiconductor containing an azo-linked carbon nitride framework for photocatalytic, photovoltaic and adsorbent applications. *J Am Chem Soc* 2019; 141(13): 5415–36.
95. Cai W, Li K, Zhu Z, et al., Construction of microreactor-like square-hole carbon nitride boosting photocatalytic nitrogen fixation activity. *Ind Eng Chem Res* 2024; 63(29): 12852–62.
96. Jiang L, Yuan X, Pan Y, et al., Doping of graphitic carbon nitride for photocatalysis: A review. *Appl Catal B* 2017; 217: 388–406.
97. Lu S, Li C, Li HH, et al, The effects of nonmetal dopants on the electronic, optical and chemical performances of monolayer g-C_3N_4 by first-principles study. *Appl Surf Sci* 2017; 392: 966–74.
98. Yang S, Wang Q, Wang Q, et al., Linkage engineering mediated carriers transfer and surface reaction over carbon nitride for enhanced photocatalytic activity. *J Mater Chem A* 2021; 9(38): 21732–40.
99. Yan SC, Li ZS, Zou ZG, Photodegradation of Rhodamine B and methyl orange over boron-doped g-C_3N_4 under visible light irradiation. *Langmuir* 2010; 26: 3894–901.
100. Fouladi Z, Saki S, Hosseini-Sarvari M, A comparative study on the photocatalytic performance of modified graphitic carbon nitride (g-C_3N_4) with non-metals (P, B, S) for aerobic oxidative desulfurization. *J Photochem Photobiol A* 2024; 453: 115652.
101. Hu S, Li F, Fan Z, et al., Band gap-tunable potassium doped graphitic carbon nitride with enhanced mineralization ability. *Dalt Trans* 2015; 44: 1084–92.
102. Vinu A, Ariga K, Mori T, et al., Preparation and characterization of well-ordered hexagonal mesoporous carbon nitride. *Adv Mater* 2005; 17: 1648–52.
103. Jun YS, Hong WH, Antonietti M, Thomas A, Mesoporous, 2D hexagonal carbon nitride and titanium nitride/carbon composites. *Adv Mater* 2009; 42: 4270–4.

104. Chen XF, Jun YS, Takanabe K, et al., Ordered mesoporous SBA-15 type graphitic carbon nitride: A semiconductor host structure for photocatalytic hydrogen evolution with visible light. *Chem Mater Commun* 2009; 21: 4093–5.
105. Gao X, Li QY, Wang YL, et al., A facile soft-hard template cooperative organization approach for mesoporous *g*-C_3N_4 with high photocatalytic performance. *Appl Surf Sci* 2024; 657: 159574.
106. Qin RX, Liu KL, Wu QY, et al., Surface coordination chemistry of atomically dispersed metal catalysts. *Chem Rev* 2020; 120: 11810–99.

3 Hydrogen Evolution (Water Splitting)

Molecular hydrogen (H_2) is attracting attention as a renewable energy source, which is obtained via water splitting by visible-light active photocatalysts. Various photocatalysts, mostly two-dimensional materials, have been proposed, such as germanium monochalcogenide (GeS and GeSe) [1], multilayer M_2CO_2 (M = Sc, Zr, Hf) [2], and $ZnIn_2S_4$-coated carbon spheres [3]. Recently, metal-free graphitic carbon nitride (g-C_3N_4)-based materials have gained increasing interest for overall water splitting, due to their structural stability, low cost, easy synthesis, environmental friendliness, and numerous efforts to enhance water splitting efficiency. It was reviewed by focusing on carbon and nitrogen vacancies in g-C_3N_4 (g-CN) [4].

In this chapter, the recent results on H_2 evolution via water splitting over g-C_3N_4-based photocatalysts, including modified and composited materials, are summarized. In addition, a recently developed catalyst, C_3N_5-based materials, is explained.

3.1 g-C_3N_4 AND MODIFIED g-C_3N_4

A polymeric semiconducting *g*-CN has attracted attention as a metal-free visible-light photocatalyst for water splitting [5]. It was synthesized by heating either cyanamide (CN_2H_2) in air at 550°C for 4 h [5] or melamine ($C_3H_6N_6$) at 500°C for 4 h in N_2 [6]. Tailoring the molecular structure of the *g*-CN is challenging to enhance photocatalytic activity through enhancements in carrier transfer efficiency and surface activation. Photocatalysts for H_2 evolution via water splitting were reviewed by focusing on two-dimensional carbon-based semiconductors, including *g*-CN [7].

The H_2-evolution rate could be enhanced by repeated calcination of *g*-CN synthesized from dicyandiamide (DCDA) by calcination at 550°C for 3 h [8]. The pristine *g*-CN was calcined repeatedly at 600°C for 2 h up to 7 times. The H_2-evolution rate increased from 31.3 μmol/h·g for the pristine *g*-CN to 764.8 μmol/h·g after the seventh reheating, although the crystallinity improvement was detected in a shift of the 002 diffraction peak from 27.26° to 27.58° in 2θ CuKα.

g-CN was synthesized from three precursors, urea, DCDA, and thiourea, by heating at 600°C in air, of which H_2-evolution performance after loading 3 wt% Pt (co-catalyst) is compared in Figure 3.1a and b under UV-vis light irradiation, respectively [9]. The urea-derived *g*-CN delivered a high H_2 evolution, much higher than those derived from thiourea and DCDA. The H_2-evolution rate was calculated to be 3,300 μmol/h·g for the urea-derived *g*-CN, more than 10 times faster than the DCDA-derived *g*-CN (300 μmol/h·g) and nearly 7 times faster than the thiourea-derived *g*-CN (500 μmol/h·g). This difference in the H_2 evolution activity observed on the photocatalysts synthesized from different precursors could not be

DOI: 10.1201/9781003589051-3

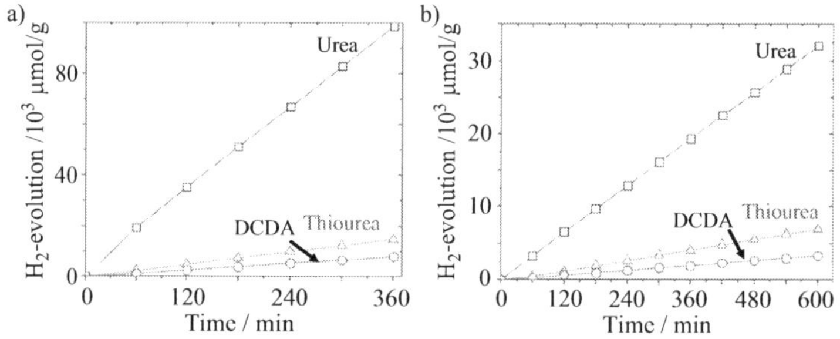

FIGURE 3.1 H_2 evolution using *g*-CN photocatalysts synthesized from urea, thiourea, and dicyandiamide (DCDA) after 3 wt% Pt-deposition in water containing triethanolamine (TEOA) at room temperature under the irradiation of a 300 W Xe lamp: (a) full arc, (b) $\lambda >$ 395 nm visible light [9].

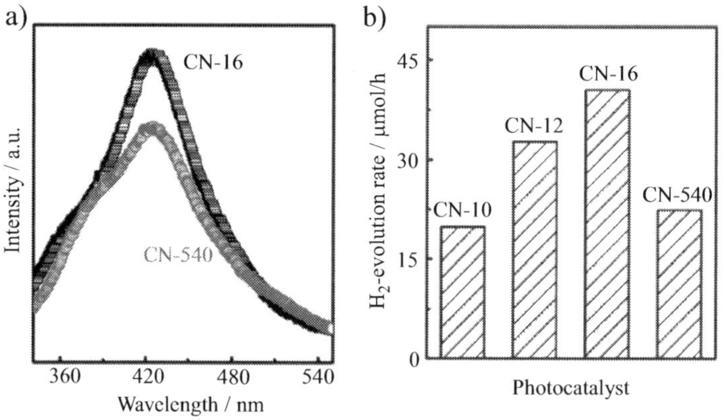

FIGURE 3.2 CNs synthesized by microwave irradiation of supramolecular aggregates for different periods of time (CN-10–CN-16) in comparison with the CN synthesized by conventional heating at 540°C (CN-540): (a) UV-vis absorption spectra and (b) H_2-evolution rate [11].

explained by the differences in surface area. The exceptionally high H_2-evolution performances of the *g*-CN derived from urea were concluded to be owing to both a high degree of polymerization and a low level of protonation, which was supposed to result from an oxygen-containing precursor. The *g*-CN synthesized from urea at 550°C for 3 h was also reported to have a high H_2-evolution rate as 590 μmol/h·g [10].

Highly crystalline *g*-CN was synthesized by microwave heating of the supramolecular aggregates derived from a mixture of melamine with cyanuric acid (MCA complex) for different periods of time (10, 12, and 16 min) [11]. The UV-vis absorption spectra and the H_2-evolution performances are compared on the CNs prepared by microwave irradiation for different times (coded as CN-10, CN-12, and CN-16, respectively) and the CN obtained by conventional furnace heating at 540°C (CN-540) in Figure 3.2a and b. Microwave-assisted CN synthesis is shown to be effective for the enhancement of H_2 evolution. By using 20 mg of catalyst under

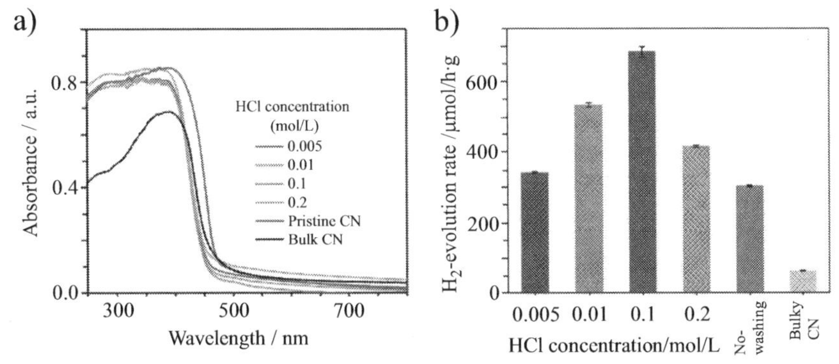

FIGURE 3.3 Highly crystalline *g*-CN synthesized from the mixture of melamine and KCl/LiCl at 550°C: (a) UV-vis absorption spectra and (b) effect of HCl concentration for washing on H_2-evolution rate [12].

visible light irradiation, the catalyst CN-16 could produce about 400 μmol H_2, while CN-540 produced about 200 μmol.

The highly crystalline *g*-CN synthesized from melamine using KCl/LiCl eutectic molten salt exhibited a high H_2-evolution performance [12]. The resultant *g*-CN by heating a mixture of melamine and KCl/LiCl at 550°C for 4 h was washed with HCl solutions with different concentrations (0.005, 0.01, 0.1, and 0.2 mol/L). The effects of HCl-concentration for washing on UV-vis absorption spectra and H_2-evolution rate measured over the resultant *g*-CNs under visible light irradiation are compared in Figure 3.3a and b, respectively. UV-vis spectra could not differentiate by HCl-concentration, while H_2-evolution rate depends strongly on HCl-concentration. XRD did not suggest a certain effect of HCl-concentration, in other words, no difference in crystallinity of *g*-CN, evaluated by diffraction angle and broadness of the 002 diffraction peak, after washing with HCl of different concentrations. Therefore, a marked difference in H_2-evolution rate shown in Figure 3.3b might be due to residual salts in the *g*-CNs after washing. Highly crystalline *g*-CN was synthesized by mixing the melamine-derived *g*-CN (600 mg) with KCl (3.3 g) and LiCl (2.7 g) in a glovebox and heating at 550°C for 4 h in N_2 (coded as CN-1) [13]. For comparison, the melamine-derived *g*-CN was directly heated at 550°C for 4 h without the KCl/LiCl molten salt (coded as CN-2), and Li^+ and Cl^--doped PTI (poly-triazine imide) was synthesized by heating a mixture of melamine (1.2 g) with KCl (6.6 g) and LiCl (5.4 g) at 450°C and then heated at 550°C in a vacuum for 24 h (coded as Li^+,Cl^--PTI). The H_2-evolution rate was measured by dispersing 50 mg of catalysts after 3 wt% Pt was deposited in 100 mL deionized water at room temperature under irradiation of light from 300 W Xe lamp. The H_2-evolution rate was calculated as 27.5 μmol/h for CN-1, 0.3 μmol/h for the CN-2 and 0.1 μmol/h for Li^+,Cl^--PTI, suggesting the importance of the crystallinity of the catalyst *g*-CN.

A crystalline *g*-CN with triazine/heptazine (PTI/PHI) heterojunctions exhibited a high H_2-evolution performance [14]. It was prepared by mixing *g*-CN nanotubes (*nt*-CN) with KCl/LiCl and then heating at 550°C for 4 h in N_2 flow to obtain rod-arrayed CN particles with heterojunctions (code as the rod-arrayed CN). *nt*-CN was synthesized by hydrothermal treatment of melamine aqueous solution at 200°C for 12 h. For comparison, bulk *g*-CN was prepared by calcination of melamine at 550°C for 4 h, poly(triazine imide) (PTI) was synthesized by heating melamine with LiCl/KCl at 400°C for 4 h and then reheating in a sealed glass tube at 550°C for 12 h, and poly(heptazine imide) (PHI) was synthesized by heating a mixture of 1 g melamine-derived *g*-CN (melon) with 2 g potassium thiocyanate (KSCN) at 400°C for 1 h and then reheating at 500°C for 4 h. H_2 evolution was measured by dispersing the catalysts (20 mg) in 40 mL aqueous solution containing 10 mL triethanolamine (TEOA) with 4 wt% Pt under the light from a 300 W Xe lamp with a 420 nm filter. The rod-arrayed CN catalyst exhibited the highest H_2-evolution rate of 11.720 mmol/h·g, which was supposed to be attributed to the presence of the triazine-heptazine (PTI-PHI) heterojunctions. The *nt*-CN gave 3.690 mmol/h·g, while the PTI gave poor activity of 0.215 mmol/h·g.

A crystalline *g*-CN was synthesized by calcination of the mixture of melamine (3 g) with KCl (2 g) prepared in ethanol at 550°C for 3 h [15]. The added KCl (melting point of 776°C) worked as a template, supplying K^+ and Cl^--intercalates. The resultant K^+ and Cl^- intercalated CN had a bandgap of 2.71 eV, a little smaller than the pristine CN (2.78 eV) and a high H_2-evolution performance, delivering a high rate of 1,356 μmol/h·g with sufficient cyclability, about 22 times higher than that of 60 μmol/h·g of the pristine CN.

A crystalline CN nanotube (*nt*-CN) was synthesized using a eutectic mixture of LiCl·H_2O/KCl/NaCl (1/1/1 by weight, having a low melting point of 355°C) [16]. Melamine was mixed with the molten salt in an salt/melamine weight ratio of 15/1, which was heated at 500°C for 1 h under semi-closed environment and then washed thoroughly with deionized water several times to obtain the intercalated *nt*-CN, of which XRD pattern is shown in Figure 3.4a, revealing a series of 00*l* diffraction peaks, as characteristics of intercalation compounds, of which intercalates are supposed to be Cl^-, K^+, and Na^+. Its chemical composition was determined to be $C_3N_{4.52}H_{3.53}K_{0.32}Na_{0.12}Li_{0.25}Cl_{0.08}$. Its lattice parameters were calculated to be $a_0 = 0.7134$ and $c_0 = 0.6634$ nm with fifth-stage intercalation structure. Its H_2-evolution performances were measured by dispersing 0.1 g catalyst in either an aqueous solution (100 mL) containing 10 vol% TEOA as the sacrificial electron donor or a saturated aqueous solution of KCl, NaCl, and LiCl with 10 vol% TEOA under visible light ($\lambda > 420$ nm)

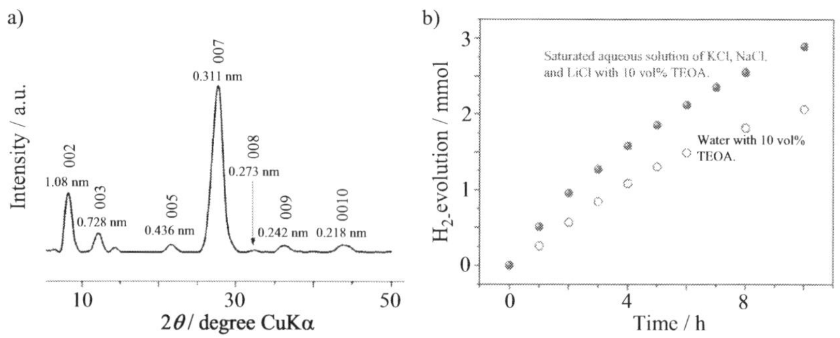

FIGURE 3.4 Crystalline *g*-CN synthesized in LiCl·H_2O/KCl/NaCl eutectic salt (fifth-stage intercalated CN): (a) XRD pattern and (b) H_2-evolution kinetics in two reaction solutions [16].

and 3 wt% Pt (co-catalyst) photo-deposited on the catalyst by using H_2PtCl_6 dissolved in the reaction solution. H_2-evolution kinetics over the resultant catalyst in two reaction solutions are shown in Figure 3.4b. The H_2-evolution rate in the first-hour reaction reaches 0.346 mmol/h, being 34.6 times higher than the pristine CN (0.01 mmol/h).

The porous, distorted *g*-CN was synthesized via a two-step process [17]. First, melamine (2 g) was added to 50 mL of HNO_3 with stirring (called N-melamine), which was dispersed into a water/ethanol mixture (1/1 by volume) with cyanuric acid (CA) and stirred until the complete evaporation of the solvent (water and ethanol). The resultant supramolecular assembly was calcined at 550°C and coded as AC-CN, in which the tri-*s*-triazine ring is distorted due to the replacement of the N-atom by an O-atom. For comparison, the *g*-CN was prepared from melamine, and a mixture of CA with melamine by calcining at 550°C, being coded as A-CN and C-CN, respectively. The photocatalytic performance of the resultant catalysts is assessed using TEOA sacrificial agent and 1 wt% Pt co-catalyst. In Figure 3.5a, the photoluminescence spectra are shown on the resultant catalysts, the emission peak intensity of AC-CN is lower than that of CN, A-CN, and C-CN, suggesting a reduced carrier recombination. In Figure 3.5b and c, H_2-evolution kinetics and evolution rates are shown, respectively, revealing a marked enhancement in H_2 evolution over the AC-CN catalyst, evolution rate reaching 7.46 mmol/g·h. It exhibited stable cyclability for H_2 evolution.

H_2-evolution performance was compared among macroscopic assemblies of low-dimensional *g*-CNs synthesized from the mixtures of melamine with cyanuric acid (MCA) and with trithiocyanuric acid (MTCA) using aprotic DMSO and protic H_2O as solvents with different precipitation temperatures, 30°C–150°C and calcined at a temperature of 500°C–600°C [18]. The resultant *g*-CNs coded by the precursor (MCA or MTCA), solvent (DMSO or H_2O), precipitation temperature (30°C or 150°C), and heat-treatment temperature (500°C, 550°C, and 600°C), such as MCA-DMSO-30-600. The H_2-evolution performances were studied by depositing 3 wt% Pt as a co-catalyst and in water with 10 vol% TEOA as an electron donor under visible light irradiation ($\lambda > 420$ nm). H_2-evolution rates of the resultant *g*-CNs are shown in Figure 3.6a, revealing that MCA and MTCA-derived *g*-CN gave a significant enhancement in H_2-evolution rate by a factor of 4–9 over the pristine melamine-derived one (Melamine-550). Cyclability for H_2 evolution was confirmed on all resultant photocatalysts, as shown in Figure 3.6b.

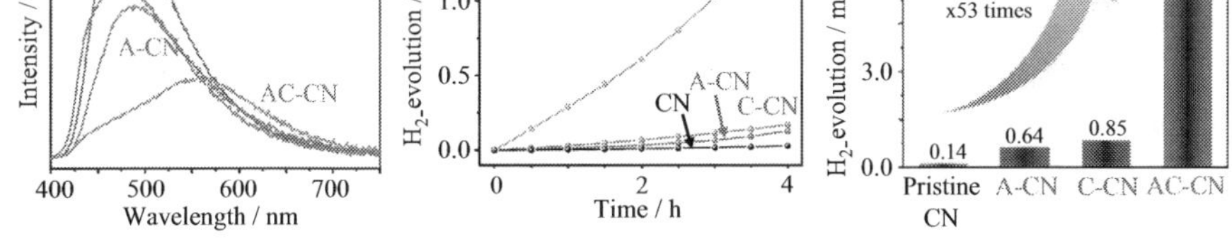

FIGURE 3.5 The porous distorted *g*-CNs, AC-CN, A-CN. C-CN and the pristine CN: (a) UV-vis absorption spectra, (b) H_2-evolution kinetics, and (c) H_2-evolution rate [17].

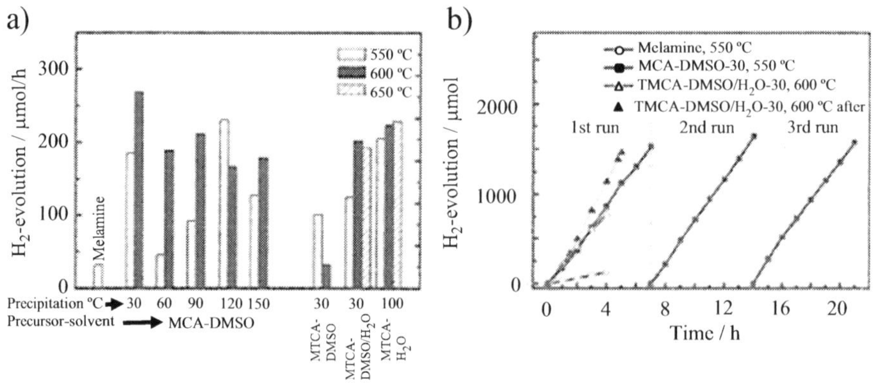

FIGURE 3.6 Photocatalytic activity of *g*-CN synthesized from the precipitates of melamine with cyanuric acid (MCA) and with trithiocyanuric acid (MTCA) in DMSO and H_2O at different precipitation temperatures of 30–150°C and by calcination at 500°C, 550°C, and 600°C: (a) H_2 evolution volume and (b) cycle stability for H_2 evolution [18].

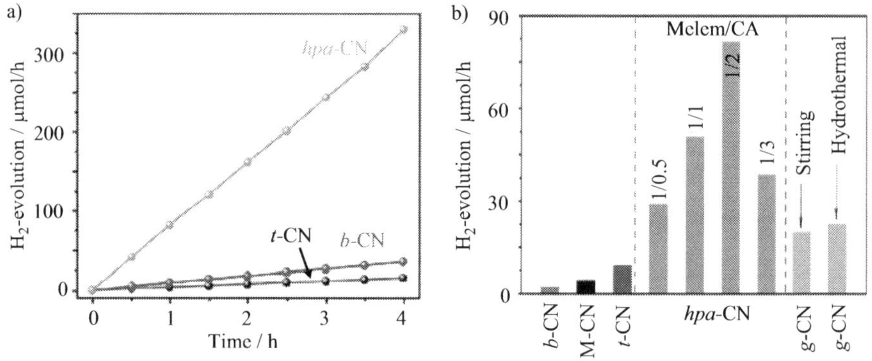

FIGURE 3.7 Hierarchically porous amorphous CN (*hpa*-CN) in comparison with *b*-CN, M-CN, and *t*-CN: (a) H_2-evolution kinetics and (b) comparison of H_2-evolution rate on different CNs [19].

Hierarchically porous amorphous CN (*hpa*-CN) prepared by mixing melem with cyanuric acid (CA) in different molar ratios of 1/0.5, 1/1, 1/2, and 1/3 in deionized water, hydrothermally treating at 180°C for 8 h, and by freeze-drying under vacuum [19]. The melem-derived *g*-CN (M-CN) was also synthesized through repolymerization of melem at 550°C for 2 h without adding CA. The melem was synthesized from melamine at 425°C. For comparison, a bulky *b*-CN was synthesized through re-polymerization of melem at 550°C for 2 h and a tubular *g*-CN (*t*-CN) was prepared through the same procedure as *hpa*-CN but without adding CA. The time change in H_2 evolution is compared with the resultant CNs, i.e., *hpa*-CN, *b*-CN, and *t*-CN, in Figure 3.7a, revealing that the *hpa*-CN with a hierarchical pore structure is effective for the enhancement of H_2-evolution rate. In Figure 3.7b, H_2-evolution rates are compared on *g*-CNs prepared in different morphologies (bulky and tubular), precursor

TABLE 3.1
Pt-Loaded *g*-CN with Different *C/N* Ratios [20]

C/N Atomic Ratio	BA/DDA Ratio in Reaction	BET Surface Area (m^2/g)	Bandgap (eV)	H_2-Evolution Rate (mmol/h)	
				λ>300 nm	λ>420 nm
0.73	0.00	10	2.67	148.2	6.5
0.74	0.05	22	2.57	251.1	29.4
0.75	0.1	19	2.45	218.8	16.4
0.76	0.2	14	2.26	175.7	12.7
0.80	0.5	12	1.86	115.3	8.9
0.88	1.0	11	1.68	53.9	5.1
0.96	2.0	12	1.58	8.5	1.6

TABLE 3.2
H_2-Evolution Rate of *mpg*-CN Loaded with Pt (3 wt%) [22]

Photocatalyst	Template Mass Ratio[a]	BET Surface Area (m^2/g)	H_2-Evolution Rate (μmol/h)
g-C_3N_4	–	8	18
mpg-C_3N_4	0.2	67	149
	0.5	126	142
	1.0	235	124
	1.5	373	69
g-C_3N_4	NH_4NF_3-washed	10	19

[a] Mixing of cyanamide with silica in a cyanamide/silica mass ratio.

mixing (different melem/CA ratios), and re-polymerization procedure (stirring and hydrothermal), revealing that a mixing of precursors (melem and CA) is markedly effective, particularly the 1/2 ratio being remarkable.

g-CNs with different *C/N* atomic ratios were synthesized via co-polymerization of the mixtures of DCDA with barbituric acid ($C_4H_4N_2O_3$, BA) in different weight ratios of 0.05, 0.1, 0.2, 0.5, 1, and 2 in aqueous solution at 550°C [20]. The H_2-evolution performances were studied on the water/TEOA mixtures under UV-vis light irradiation. The results are listed together with bandgap in Table 3.1, revealing that the resultant catalysts are active under visible light and even under UV light. Visible light activity was very sensitive to the *C/N* ratio.

Mesoporous *g*-CN (*mpg*-CN), which was prepared by using a mesoporous silica template, exhibited a marked accelerated H_2-evolution rate [21]. The H_2-evolution rates of these *mpg*-CNs are summarized in Table 3.2, which were measured by dispersing the catalyst powder of 0.1 g after Pt deposition in an aqueous solution (100 mL) containing TEOA (10 vol.%) as sacrificial electron donor [22]. Table 3.2 shows a marked enhancement in H_2-evolution rate on *mpg*-CNs. A similar effect of

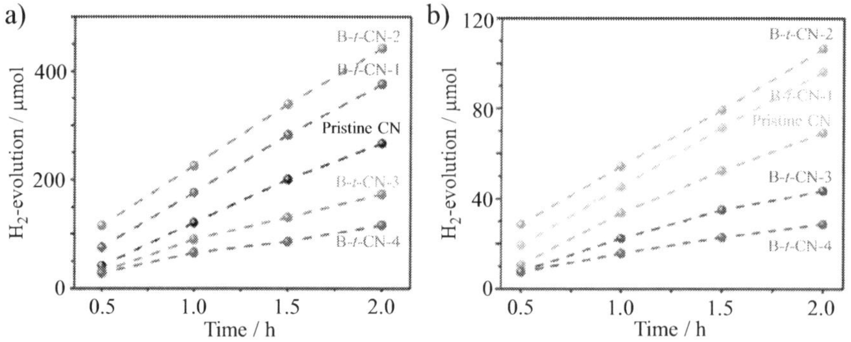

FIGURE 3.8 H_2-evolution performances of B-*t*-CNs with different B-dosages: (a) under simulated sunlight and (b) under 36 W LED lamps [25].

the introduction of mesopores into bulk *g*-CN was reported on the *mpg*-CN prepared using tetraethyl orthosilicate (TEOS) as template precursor [23] and ordered mesoporous silica SBA-15 [24].

Doping of hetero-atoms and hetero-molecules into the *g*-CN framework was reported as an effective structure modification for improving the photocatalytic activity of *g*-CN, including H_2 evolution.

B-doping of *g*-CN was shown to be effective in enhancing H_2 evolution [25]. B-doped tubular CNs (B-*t*-CNs) with different B-dosages were prepared by heating the mixtures of *g*-CN with boric acid (H_3BO_3) in different ratios (0.025, 0.05, 0.10, and 0.20) at 550°C. The process yielded B-contents of 1.71, 2.67. 5.04, and 8.26 at%, respectively (coded as B-*t*-CN-1, -2, -3, and -4, respectively). H_2-evolution performances of the pristine CN and the B-*t*-CNs are shown under the irradiation of a simulated sunlight (Figure 3.8a) and under 36 W LED lamp (Figure 3.8b). The H_2-evolution rates under two irradiation conditions were calculated to be 121.30 and 34.62 µmol/h for the pristine CN, 189.95 and 48.22 µmol/h for the B-*t*-CN-1, 220.80 and 53.21 µmol/h for the B-*t*-CN-2, 86.70 and 21.79 µmol/h for the B-*t*-CN-3, and 58.45 and 14.43 µmol/h for the B-*t*-CN-4, respectively. The highest H_2-evolution performance was obtained by the B-*t*-CN-2 catalyst, being approximately 1.82 and 1.54 times higher than that of the pristine CN nanosheets under simulated sunlight and LED lamp irradiation, respectively. It was confirmed to have a high cycle stability for H_2 evolution. The bandgaps of the pristine CN and B-*t*-CNs progressively decreased from 2.63 eV for the CN to 2.47 eV for the *t*-B-CN-2 with increasing B-dosage.

C-doped self-assembled supramolecular CNs (*sa*-C-CNs) synthesized from a mixture of thiourea and trimesic acid ($C_9H_6O_6$), which is composed of wrinkled sheets, demonstrated a high H_2-evolution rate of 1,405 µmol/h·g, being 4.4 times higher than that of the pristine *g*-CN [26].

C and O co-doping of *g*-CNs (C,O-CNs) were prepared by mixing CA/melamine (20 mmol, each) with different amounts of benzonitrile (0, 50, 75, 100, and 125 µL) in water under stirring, following drying at 60°C and annealing at 550°C for 4 h, coded as C,O-CN-0, -1, -2, -3, and -4, respectively [27]. Based on the data measured by various characterization techniques, such as XPS, XRD, and FTIR, the doped O and

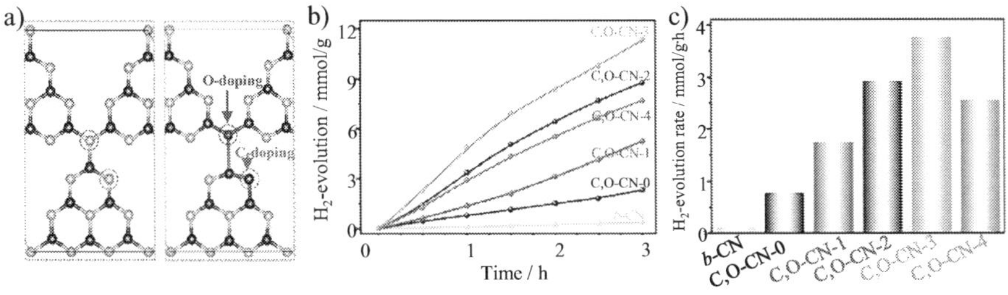

FIGURE 3.9 C and O-co-doped *g*-CNs (C,O-CN-1–C,O-CN-4) in comparison with B-CN: (a) O and C doped positions, (b) H_2-evolution kinetics, and (c) H_2-evolution rates [27].

C atoms were supposed to replace the N atoms in *g*-CN as illustrated in Figure 3.9a. H_2-evolution performances over the resultant C,O-CNs were measured under the irradiation of a visible light ($\lambda \geq 420$ nm) from a 300 W Xe lamp. The results are shown in comparison with bulk *g*-CN derived from melamine (*b*-CN) in Figure 3.9b and c, revealing that the C,O-CN-3 catalyst has the highest H_2-evolution performances, much higher than *b*-CN and that the effectiveness of C and O co-doping into the triazine framework.

The construction of a CN composed of melem units linked by a C–O–C chain (O-linked CN) was prepared via copolymerizing the melem with formaldehyde at 450°C, 500°C, and 550°C for 4 h in N_2 (coded as O-linked CN-1, -2, and -3, respectively) [6]. The resultant O-linked CNs had a columnar morphology with many crack-like spaces. H_2-evolution performances of O-linked CNs were measured using 20 mg of catalyst after Pt-deposition in a 40 mL water containing 10 mL of TEOA (a sacrificial reagent), 250 μL of H_2PtCl_6 under visible light irradiation. The introduction of O-ligands significantly enhanced H_2 evolution, of which rate is accelerated with the increase in their doping concentration. An additional annealing at 500°C (O-linked CN-2) gave the highest H_2 evolution with stable cycle performance. The evolution rates for 1 g catalysts of the pristine CN, O-linked CN-1, -2, and -3 were estimated to be 0.280, 1.230, 1.745, and 0.795 mmol/g·h, and their bandgaps were calculated to be 2.58, 2.60, 2.52, and 2.48 eV, respectively.

Atom P was doped to link the melem units by a C–P–C chain, which was also effective to enhance H_2 evolution [28]. P-doping was performed by calcining a mixture of melem with phosphonitrilic chloride trimer (PCT, $Cl_6N_3P_3$) in a molar ratio of 1/1 at 500°C, 550°C, 600°C, and 650°C in N_2. The resultant P-doped CNs are coded as P-CN-*X*, where *X* corresponds to heat treatment temperature. For comparison, *g*-CN was synthesized by calcination of melem at 500°C for 4 h (coded as CN) and by calcination of a mixture of melamine with PCT (molar ratio 1/1) at 400°C (coded as MP-CN). The H_2-evolution performances of these photocatalysts were measured by dispersing the catalyst (20 mg) into a deionized water (50 mL) containing 10 mL of benzylamine. The H_2-evolution performances of the catalyst are shown in Figure 3.10. The references (the pristine CN and MP-CN) exhibited similar performances (relatively low evolution volume and rate). The experimental results demonstrate that the start from the mixture of melem with PCT and the calcination at 600°C are key for getting high H_2 evolution.

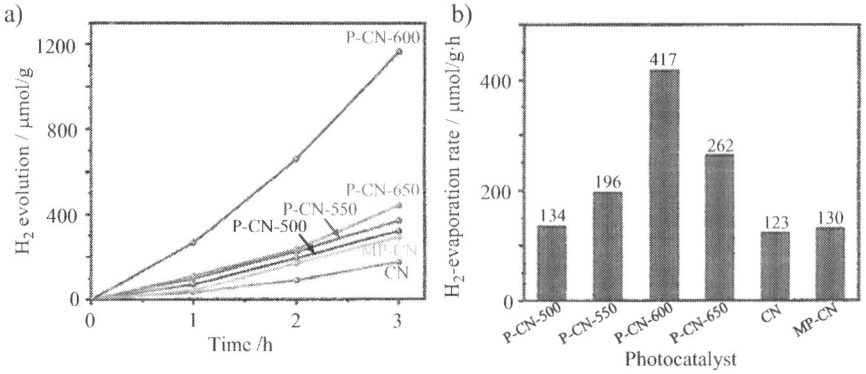

FIGURE 3.10 H_2-evolution performances of P-doped *g*-CNs (P-CN-500–650) in comparison with the CN and MP-CN: (a) time change of H_2 evolution and (b) H_2-evolution rate [28].

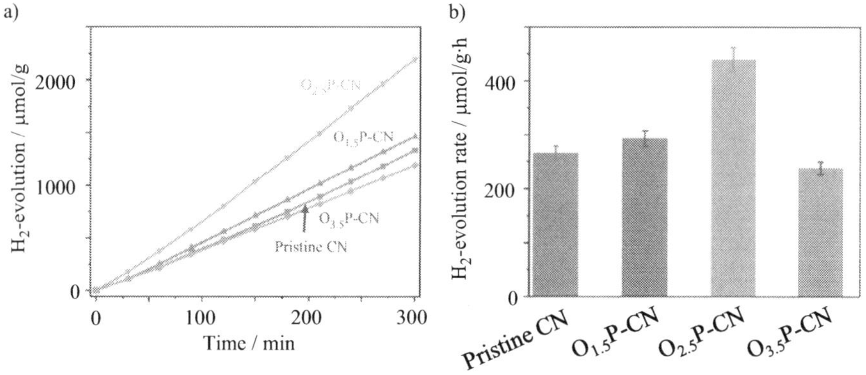

FIGURE 3.11 H_2-evolution performances of the O and P co-doped O_XP-CNs: (a) evolution kinetics and (b) evolution rate [29].

The O and P co-doped *g*-CNs were synthesized using the one-step thermal polymerization [29]. The precursor mixtures of urea (10 g) with a fixed amount of H_3PO_2 (P-source), and different amounts of ammonium persulfate $(NH_4)_2S_2O_8$ (O-source) (1.5, 2.5, and 3.5 g) were calcined at 550°C for 2 h. The resultant catalysts were denoted as $O_{1.5}$P-CN, $O_{2.5}$P-CN, and $O_{3.5}$P-CN, respectively. For comparison, the pristine *g*-CN was synthesized by the same procedure without adding P- and O-sources. On the resultant catalysts after about 3 wt% Pt deposition, H_2 evolution was measured by dispersing 30 mg catalyst powders in 100 mL water containing 10 mL TEOA under the solar simulated light (400–1,100 nm wavelength and 100 mW/cm^2 intensity) from a diode lamp. H_2-evolution kinetics and evolution rate are compared on the resultant catalysts in Figure 3.11a and b, respectively. The $O_{2.5}$P-CN exhibited the highest H_2 evolution of 2,198 μmol/g, whereas $O_{3.5}$P-CN gave even lower evolution than the pristine CN (1,330 μmol/g). The highest H_2-evolution rate reached 439.7 μmol/g·h for $O_{2.5}$P-CN and the apparent quantum efficiency was calculated to be 7.33%.

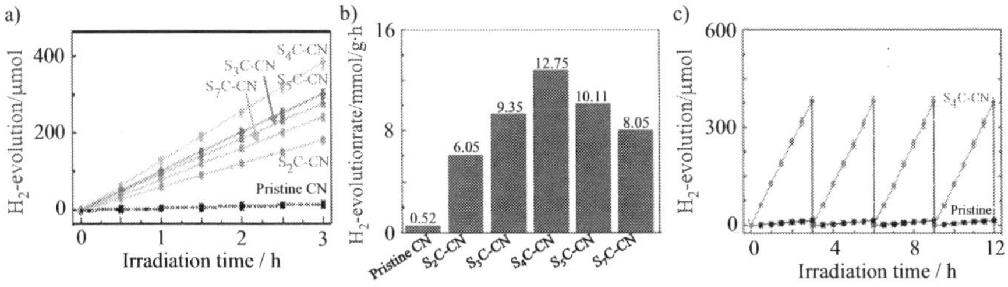

FIGURE 3.12 H_2-evolution performances of S and C co-doped CN catalysts (S_XC = CN, X = 2, 3, 4. 5, and 7) in comparison with the pristine CN: (a) H_2-evolution kinetics, (b) H_2-evolution rate, and (c) cyclabilities of the pristine CN and the S_4C-CN [30].

S and C co-doping of *g*-CN was reported to be effective to enhance the photocatalytic H_2 evolution [30]. The catalysts were synthesized by hydrothermal heating at 180°C on the mixtures of melamine with thiolactic acid ($C_3H_6O_2S$) by molar ratio of thiolactic acid to melamine of 0, 2, 3, 4, 5, and 7, and then the precipitates were calcined at 520°C 2 h in Ar to get S, C co-doped *g*-CN (coded as S_XC-CN, where X is the mixed molar ratio of thiolactic acid to melamine). The resultant catalysts were ultra-thin-walled tubes. After Pt photo-deposition, their H_2-evolution performances were studied in a water containing 20 vol% TEOA under the irradiation of the light from a 300 W Xe lamp with a UV (λ > 420 nm) cutoff filter at an intensity of 600 W/cm^2. The results are shown in Figure 3.12a and b by comparing with the pristine *g*-CN, showing a marked enhancement effect of S and C co-doping, the S_4C-CN catalyst delivering the highest evolution rate of 12.75 mmol/g·h. The S_4C-CN exhibited highly stable cyclability with much higher H_2 evolution than pristine CN, as shown in Figure 3.12c.

g-CN derived from urea was dispersed in THF and then solvothermal treatment at 180°C for 12 h to produce an exfoliated ultrathin CN sheet (*ut*-CN) [31]. The resultant *ut*-CN was heated to temperatures of 200°C, 300°C, 400°C, and 550°C for 4 h to synthesize CNs with 2C-coordinated N-vacancy (V_{N2C}) (V_{N2C}-*ut*-CN-X, X corresponds to heating temperature). Using the same procedure for *ut*-CN synthesis by adding different amounts of ammonium acetate (0.1, 0.3, and 0.5 g), *ut*-CNs with 3C-coordinated N-vacancy (V_{N3C}) (V_{N3C}-*ut*-CN-X, where X corresponds to the amount of ammonium acetate used) were obtained. On these two kinds of *ut*-CNs, Pt single atoms were loaded via in situ photo-deposition techniques by dispersing 10 mg of *ut*-CN powders in an aqueous solution of H_2PtCl_6 (10 mg/mL) under the light from an Xe lamp for 1 h. H_2-evolution performances were measured using water containing 20 vol% TEOA under light (λ>395 nm). In Figure 3.13a, H_2-evolution rate was compared among three Pt-loaded catalysts, the pristine *ut*-CN (Pt/*ut*-CN), Pt/V_{N2C}-*ut*-CN, and Pt/V_{N3C}-*ut*-CN, revealing a remarkable enhancement in H_2-evolution rate on Pt/V_{N2C}-*ut*-CN to 1.74 mmol/g·h. In contrast, Pt/V_{N3C}-*ut*-CN gave 0.92 mmol/g·h, higher than the pristine Pt/*ut*-CN. In Figure 3.13b the time-dependent photocatalytic H_2 evolution is shown, Pt/V_{N2C}-*ut*-CN exhibited a faster photocatalytic H_2-evolution rate compared to Pt/V_{N3C}-*ut*-CN, keeping excellent stability under 3 h of irradiation.

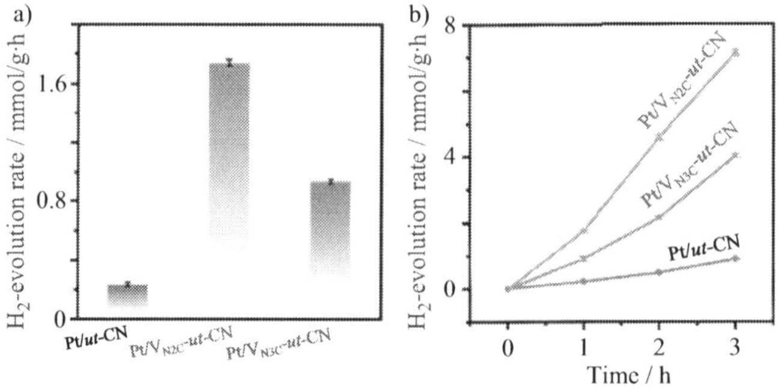

FIGURE 3.13 H_2-evolution performances of CN with 2C-coordinated N-vacancy (V_{N2C}-*ut*-CN-*X*) and CN with 3C-coordinated N-vacancy (V_{N3C}-*ut*-CN-*X*) in comparison with the pristine CN: (a) H_2-evolution and (b) evolution rate [31].

3.2 *g*-C_3N_4-BASED COMPOSITES

A self-composite CN-600/CN-500 having homojunctions was prepared in two steps, first, CN-600 was synthesized from the precursor DCDA at 600°C for 4 h, and in the second, the mixture of CN-600 with DCDA in different weight ratios DCDA/CN-600 (2/5, 4/5, and 6/5) was heated at 500°C for 4 h to obtain the CN-600/CN-500 self-composites (coded as CN/CN-1, -2, and -3, respectively, containing increasing homojunctions) [32]. After photo-deposition of 2 wt% Pt, the H_2 evolution behaviors of the resultant self-composites were studied by dispersing 100 mg of the photocatalyst in 90 mL of water with 10 mL of TEOA. H_2-evolution performances are shown in Figure 3.14a, revealing a marked enhancement of H_2 evolution for the CN/CN-1–CN/CN-3 in comparison with the pristine CNs (CN-500, CN-550, and CN-600), in other words, the introduction of homojunctions into the catalysts is very effective for enhancing H_2 evolution; the evolution rate 16.63 μmol/h for 5 mg of CN/CN-2 catalyst (corresponding to about 0.33 mmol/g·h for the rate per 1 g catalyst). The stability

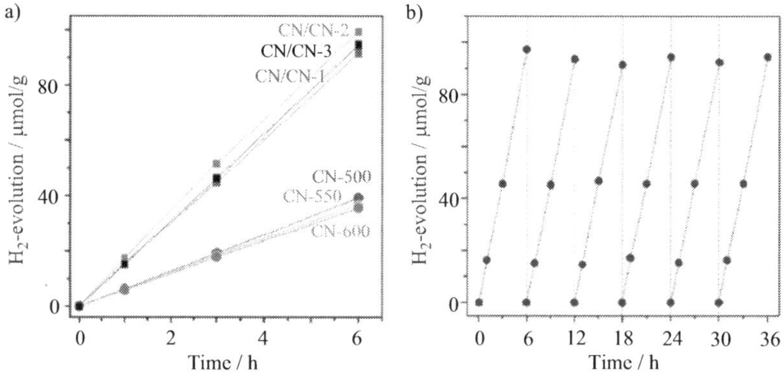

FIGURE 3.14 H_2-evolution performances of *g*-CN-based photocatalysts introducing homojunctions of CN-600/CN-500 in comparison with the pristine CN-500, CN-600: (a) H_2-evolution kinetics and (b) cycle stability of H_2 evolution on CN/CN-2 catalyst [32].

of the optimized photocatalytic system for H_2 evolution over an extended period is shown on the self-composite CN/CN-2 in Figure 3.14b. To repeat, the catalytic cycle system was purged with Ar every 6 h, and the catalyst CN/CN-2 exhibited steady H_2-evolution performance.

A novel carbon nitride homojunction with the electron-enriched and hole-enriched *g*-CN nanosheets (Re-CN and Ox-CN, respectively) was proposed to achieve high water splitting performances [33]. A layer-by-layer composite composed of Re-CN/CN/Ox-CN was fabricated by spray calcination. H_2- and O_2-evolution were measured by dispersing 10 mg catalyst powder in water containing a scavenger over seven kinds of catalysts, i.e., the pristine CN, Re-CN, and Ox-CN, mechanical mixtures of Re-CN with CN, and Ox-CN with CN, and the double-layered composites of Re-CN/CN, Ox-CN/CN, and Re-CN/Ox-CN, in addition to the triple-layered composite Re-CN/CN/Ox-CN. H_2-evolution performance was measured by dispersing the photocatalyst powder (10 mg) after deposition of 6 wt% Pt (co-catalyst) in 100 mL water containing 10 vol% TEOA. The results are shown in Figure 3.15a, revealing a marked enhancement by constructing the triple-layered composite, H_2 evolution reached 1,270.53 μmol/h (1.27 mmol/h·g), about 41 times larger than the pristine CN, and apparent quantum yield (AQY) was 71%. The O_2-evolution was measured by dispersing the photocatalyst powder of 10 mg after loading 3 wt% $Co(OH)_2$ as a co-catalyst in 100 mL water containing 5 mmol/L $AgNO_3$ as a sacrificial scavenger. The results are shown in Figure 3.15b, revealing a similar enhancement by the triple-layered composite. O_2-evolution of the composite was 36.01 μmol/h (3.60 mmol/h·g).

The self-composite B-CN/CN exhibited H_2 evolution under visible light irradiation [34]. B-doped CN (B-CN) was prepared from a mixture of urea (10 g) with Ph_4BNa (6 mg) by calcination at 550°C and B-CN/CN composites were prepared by mixing urea (10 g) with different amounts of B-CN (10, 25, and 50 mg). The photocatalytic efficiency for H_2 evolution increased with increasing B-CN-content in the composites. The photocatalytic H_2-evolution rate of $B\text{-}CN_{25}/CN$ was the highest,

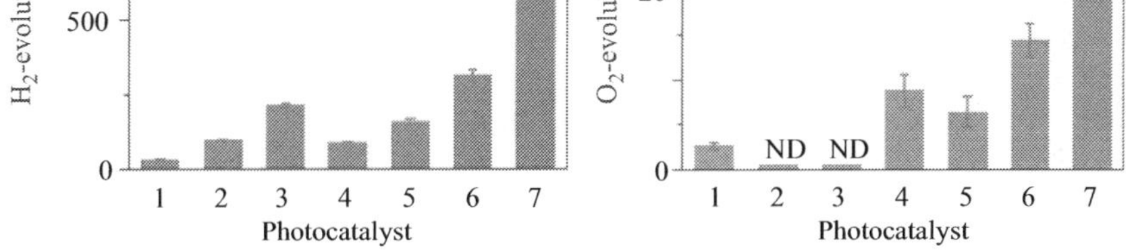

FIGURE 3.15 Water splitting over the triple-layered composite Re-CN/CN/Ox-CN in comparison with the related photocatalysts: (a) H_2-evolution using TEOA as hole scavenger and (b) O_2-evolution using $AgNO_3$ as an electron scavenger [33].

TABLE 3.3
Photocatalytic H_2-Evolution Rate of *g*-C_3N_4 Loaded with Various Metals [5]

Metal Loaded[a]	Loading Procedure	H_2-Evolution Rate (μmol/g·h)
Rh	Photo-deposition	1.6
Pd	Photo-deposition	5.7
Ir	Photo-deposition	0.2
Au	Photo-deposition	3.7
Pt	Photo-deposition	7.3
Ru	Photo-deposition	2.1

[a] 0.5 wt% metal-loaded.

being about 3.4 and about 1.8 times higher than that of the pristine CN and B-CN, respectively. The composite B-CN/CN exhibited a similar enhanced performance for photocatalytic degradation of methyl orange and phenol.

Often, nanoparticles of various metals were loaded onto the surface of *g*-CN through in situ photo-deposition in either $(NH_4)_2RuCl_6$, Na_3RhCl_6, $(NH_4)_2PdCl_4$, Na_2IrCl_6, H_2PtCl_6, or $HAuCl_4$, as a co-catalyst for photocatalytic reactions. On Pt and Ru, loading was performed by an impregnation method using THF solutions of bis(1,5-cyclooctadiene) platinum complex, $Pt(cod)_2$, and triruthenium dodecacarbonyl, $Ru_3(CO)_{12}$. In Table 3.3, H_2-evolution rates measured under the irradiation of a visible light from an Xe lamp (300 W) with a cutoff filter from a dispersion of 0.1 g *g*-CN catalyst powder loaded with various metals in a 0.01 M $AgNO_3$ aqueous solution (100 mL) containing 0.2 g of La_2O_3 [5]. The metal loading enhanced the H_2-evolution rate, depending on the metal loaded: Pt-loading was the most effective, while Ir had only a little promotional effect. Therefore, Pt-loading has been frequently employed to study the photocatalytic performances for various catalysts, as explained above on various *g*-CN-based ones.

Loading of crystalline N-doped $K_2Ta_2O_6$ (KT) and $H_2Ta_2O_6$ (HT) (coded as the composite KT/CN and HT/CN, respectively) was reported to be effective in enhancing the H_2-evolution performances [35]. In Figure 3.16a, UV-vis diffusion reflectance spectra for the composites are shown in comparison with the pristine KT and HT, revealing that a broad absorbance due to CN is added on the spectra for KT and HT on the composites, KT/CN and HT/CN. The energy bandgaps of KT, HT, KT/CN, and HT/CN were calculated to be 4.8, 4.7, 2.8, and 2.9 eV, respectively. In Figure 3.16b, time changes of the evolved H_2 from methanol/water solution using 0.3 g catalysts under visible light (λ>410 nm) are compared on the composites KT/CN and HT/CN and the pristine CN, demonstrating the effectiveness of the loading of HT and KT onto CN. Here, the CN, which was synthesized by the decomposition of the pure urea at 400°C for 2 h was used as the reference (S_{BET} of 28.0 m^2/g). The average H_2-evolution rates of KT/CN and HT/CN were 15.0 and 21.2 μmol/h, respectively (the rate calculated per 1 g of the catalysts is 50 and 70.7 μmol/g·h, respectively). The inset of Figure 3.16b shows the cycle stability of HT/CN for photocatalytic H_2 evolution.

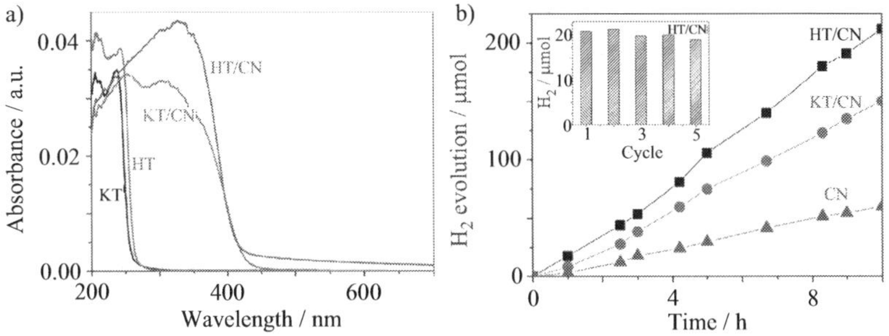

FIGURE 3.16 N-doped $K_2Ta_2O_6$ and $H_2Ta_2O_6$-loaded *g*-CN (KT/CN and HT/CN, respectively): (a) UV-vis diffusion reflectance spectra in comparison with the pristine KT and HT, and (b) H_2-evolution kinetics on Pt (0.5%)-loaded KT/CN and HT/CN catalysts in comparison with the reference CN with the inset of cycle performance of HT/CN composite [35].

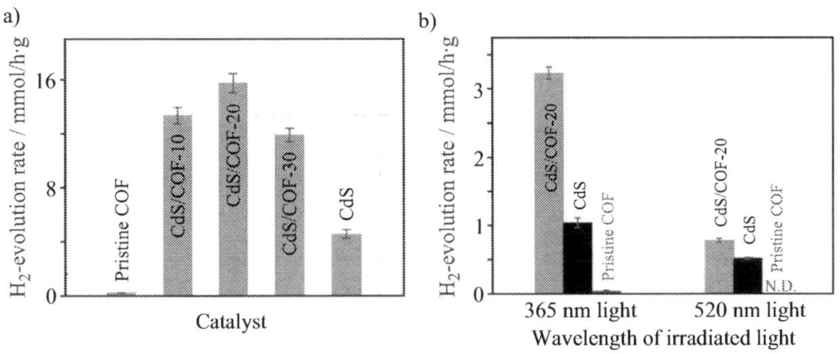

FIGURE 3.17 H_2-evolution performances of CdS-loaded COF composites (CdS/CN-*X*) in comparison with the pristine CdS and COF: (a) H_2-evolution rate and (b) effect of the wavelength of irradiated light [37].

A composite of S-doped *g*-CN (S-CN)-loaded $ZnIn_2S_4$ (ZIS) coded as S-CN_X/ZIS) exhibited an excellent H_2-evolution performance [36]. H_2-evolution rate depends strongly on the loading amounts of S-CN, 0.05, 0.1, and 0.15 g of the composites, and the H_2-evolution rate was calculated as 7.385, 8.463, and 6.735 mmol/g·h, respectively.

A triazine-based covalent organic framework (COF) was synthesized by condensation reaction of precursors, 1,3,5-tris-(4-formyl-phenyl)-triazine (TFPT) and 2,5-bis-(3-hydroxypropoxy) terephthalohydrazide (DHTH) (coded as TFPT-DHTH-COF) [37]. The resultant COF (10, 20, and 30 mg) was dispersed in 10 mL deionized water with cadmium acetate dihydrate, in which 60 mL of DMF and thiourea were added, and then hydrothermally heat-treated at 160°C for 12 h to produce the composite CdS/COF-*X*, where *X* corresponds to the weight of the pristine COF used for CdS-loading. H_2-evolution performances were measured by dispersing 10 mg catalyst in 100 mL of ascorbic acid aqueous solution (0.1 M) at 20°C. The performances are shown in comparison with the pristine COF and CdS in Figure 3.17a, revealing a remarkable enhancement in H_2-evolution rate by coupling of CdS with COF. Figure 3.17b shows the effect of the wavelength of irradiated light. The composite CdS/COF-20 delivered

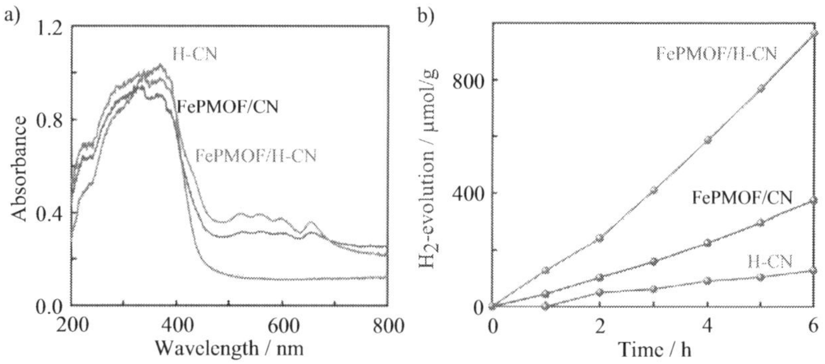

FIGURE 3.18 Photocatalytic activity of the composite FePMOF/H-CN in comparison with the pristine H-CN and FePMOF: (a) UV-vis absorbance spectra and (b) H_2-evolution kinetics [38].

the highest H_2-evolution rate of 15.75 mmol/h·g, which is about 75 times higher than that of TFPT-DHTH-COF (0.21 mmol/h·g) and 3.4 times higher than that of CdS (4.57 mmol/h·g). CdS/COF is more active under 365 nm light irradiation than under 520 nm light irradiation, as shown in Figure 3.17b.

A composite of the protonated *g*-CN loaded with FePMOF (iron porphyrin metal-organic framework) (FePMOF/H-CN) was prepared by mixing a protonated *g*-CN (H-CN) with FePMOF in deionized H_2O under sonication [38]. The H_2-evolution performances were studied by dispersing the composite powder (10 mg) in a deionized water (20 mL) containing 45 mL methanol and 5 mL TEOA under irradiation of the light from a 300 W Xe lamp. The UV-vis absorbance spectra and H_2-evolution kinetics are compared on FePMOF/H-CN composite with the pristine H-CN and FePMOF/CN in Figure 3.18a and b, respectively. The resultant composite FePMOF/H-CN has comparable absorbances to the respective components, i.e., H-CN and FePMOF, while the H_2-evolution performance of the composite is much higher than the components. The pristine H-CN shows a low photocatalytic H_2-production, while H_2 evolution increases by 7.57 times after making a composite with the FePMOF. Enhancement in H_2 evolution by the loading of FePMOF was remarkable, as shown in Figure 3.18b.

On the composites of *g*-CN with poly(3-hexylthiophene) (P3HT) (coded as P3HT/CN), XRD showed the co-existence of two phases, i.e., *g*-CN and P3HT [39]. The rate of H_2 evolution in the presence of 0.25 M Na_2S and 0.25 M Na_2SO_3 as the sacrificial agent under visible light irradiation ($\lambda > 400$ nm) is shown on the P3HT/CN composite with and without Pt loaded (1 wt%) in comparison with the pristine CN and P3HT in Figure 3.19a. In the absence of Pt, a negligible amount of H_2 was detected for both P3HT and CN. In contrast, the P3HT/CN composite, even without a Pt co-catalyst, shows an enhanced activity with a H_2-evolution rate of 1.2 μmol/h and it can be further enhanced by loading 1 wt% Pt on the composite. The effect of P3HT content in the composite is shown in Figure 3.19b, revealing a strong dependence on P3HT content. By loading 1 wt% P3HT onto CN, the H_2 evolution activity is increased by up to 90 times, and 5 wt% P3HT-loading is the optimum, reaching H_2-evolution rate of 560 μmol/h (corresponding to 1.9 mmol/g·h).

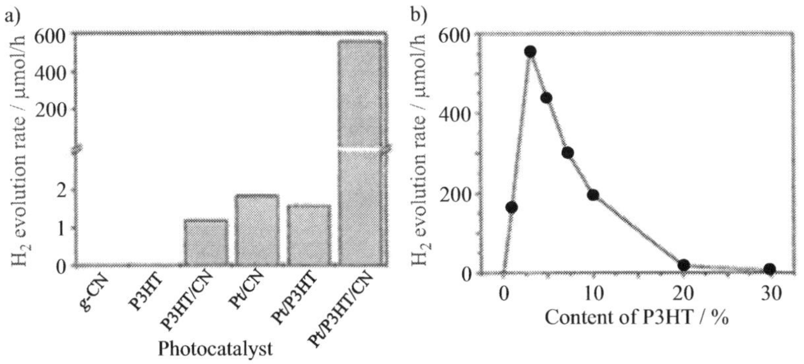

FIGURE 3.19 The H_2-evolution rate of the 0.3 g P3HT/CN composite from 0.25 M Na_2S–0.25 M Na_2SO_3 solution under visible light irradiation: (a) H_2-evolution rate of P3HT/CN composite with or without 1 wt% Pt loading in comparison with the pristine CN and P3HT, and (b) effect of P3HT content in the P3HT/CN composite on H_2-evolution rate [39].

H_2-evolution performances of three kinds of polyimide (PI) were measured in a 10 vol% methanol/water mixture [40]. Three PIs were synthesized by conjugating the melem with pyromellitic dianhydride (PMDA), 4,4-biphthalic dianhydride (BPDA), and naphthalene-1,4,5,8-tetracarboxylic dianhydride (NTDA) in equal molar ratios, coded as PI-1, PI-2, and PI-3. The S_{BET} and bandgaps of the PIs are listed in comparison with the pristine melem and *g*-CN synthesized from melamine at 550°C in Table 3.4, together with H_2-evolution rate under $\lambda > 300$ nm light and $\lambda > 420$ nm light irradiation. The H_2 evolution was measured by dispersing the 0.2 g catalyst after Pt-deposition in an aqueous solution (400 mL) containing methanol (10 vol%) as the sacrificial electron donor. As shown in Table 3.4, H_2-evolution rates of PIs were higher under $\lambda > 300$ nm light irradiation than under $\lambda > 420$ nm irradiation, while inferior to the reference *g*-CN and the pristine melem.

3.3 C_3N_5-BASED MATERIALS

The recently synthesized two-dimensional carbon nitride (C_3N_5) was predicted to suppress the recombination of the photogenerated carriers owing to its intrinsic electric field (EF) on the basis of the first-principles calculation [41]. The schemata for the non-polar photocatalysts and for the polar 2D photocatalysts with out-of-plane intrinsic EF are shown in Figure 3.20a and b, respectively. For the non-polar photocatalyst, the bandgap required for overall water splitting should be larger than 1.23 eV, because both the hydrogen reduction potential ($V_H = -4.44$ eV) and water oxidation potential ($V_O = -5.67$ eV) are located inside the bandgap, as shown in Figure 3.20a. In this case, the redox ability of the photogenerated electrons (or holes) can be evaluated by the energy difference between the CB (or VB) and the hydrogen reduction potential, which is denoted as U_e (or U_h), which correspond to the potential of the photogenerated electrons and holes, respectively. To trigger the HER and OER, it should fulfill the following conditions: U_e and U_h must be larger than the overpotentials of the HER and OER. Since the overpotentials for water splitting for an ideal photocatalyst are 0, the lower limit of bandgap (U_e+U_h) is 1.23 eV. For the polar 2D

TABLE 3.4
Physicochemical Properties and Photocatalytic H_2-Evolution Rate for the Melem-Based Polyimide Catalysts [40]

Catalyst	S_{BET} (m^2/g)	Bandgap (eV)	H_2-Evolution Rate (μmol/h)	
			$\lambda > 300$ nm	$\lambda > 420$ nm
Pristine melem	4.3	2.9	64.4	1.9
PI-1	8.6	2.5	55.0	13.3
PI-2	1.1	2.8	22.0	1.2
PI-3	5.2	1.9	10.9	1.3
Reference *g*-CN	10	2.7	93.8	7.0

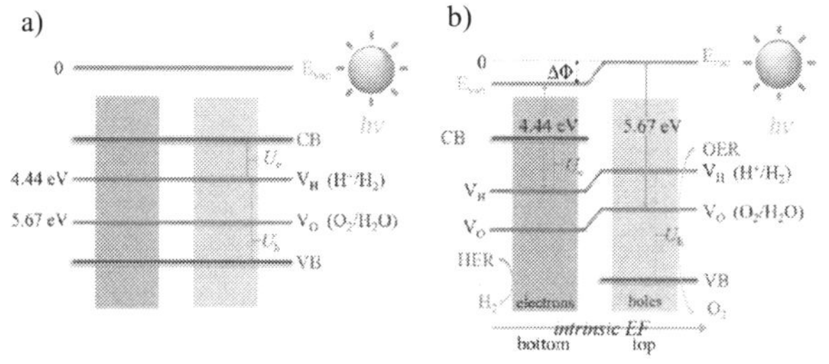

FIGURE 3.20 The schematic representations of the water splitting mechanisms for the photocatalysts: (a) without and (b) with the intrinsic EF effect [41].

photocatalysts having an out-of-plane intrinsic EF pointing from the bottom surface to the top surface, where the vacuum levels are no longer equivalent in the top and bottom regions due to intrinsic EF by $\Delta\Phi$ as illustrated in Figure 3.20b. The bandgap for water splitting over polar catalyst becomes smaller by $\Delta\Phi$ than over the non-polar catalyst. The computation predicted that the C_3N_5 multilayers can provide a bandgap tunable in a range of 2.95–2.16 eV, the broadened energy range of light absorption and lead to enhanced solar energy conversion efficiency.

3.4 H_2 PURIFICATION WITH *g*-CN

The H_2 obtained by the photocatalytic water splitting was able to be purified with the highly crystalline PTI, which is one of the polymorphs of *g*-C_3N_4.

A highly crystalline PTI was synthesized from various precursors, such as melamine, urea, using LiCl/KCl eutectic molten salt by sealing into a quartz ampule [42]. The PTI contains a high density of triangular nanopores that can accommodate a spherical guest with diameters below 0.34 nm, as illustrated its molecular structure in Figure 3.21a. This synthesis procedure is complicated and potentially unsafe due to the buildup of ammonia gas pressure (up to 12 bar), while the typical yield is only 60–80 mg of crystalline Li^+Cl^--PTI. Accordingly, the following method, which is simple and suitable for scale-up, was proposed; a PTI precursor was placed at the bottom of a crucible, which was then completely covered with the salt. The crucible

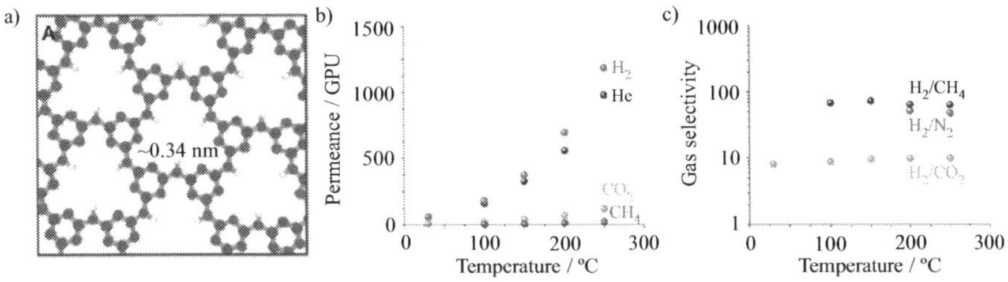

FIGURE 3.21 Poly(triazine imide, PTI): (a) molecular structure, (b) temperature dependences of permeance of different gases, and (c) temperature dependences of gas selectivity [42].

was heated to 600°C at ambient pressure, and PTI was collected by washing the salt with water with a yield of about 70%. The resultant PTI was a plate-shaped, hexagonal prismatic crystal with an average lateral size of 33 nm. By using the as-synthesized platelets as seeds for increasing the platelet size up to 95 nm. The synthesized Li^+Cl^--PTI crystal was exfoliated by stirring in anhydrous dimethylacetamide (DMAc) at 100°C for 6 h to obtain thin nanosheets with a mean thickness of 0.77 nm. The resultant exfoliated PTI nanosheets were compacted into a 900-nm-thick film by vacuum filtration. On these films of PTI nanosheets, gas-sieving performance was then evaluated. The films contain the nanopores with an effective pore aperture of 0.34 nm, exhibiting stability against harsh chemical environments and withstand temperatures up to 600°C, making them highly promising for sieving applications under extreme conditions. For example, high-temperature H_2 purification from streams containing CO_2, N_2, CH_4, and H_2S was examined. The gas permeance and gas selectivity of the PTI film for different gases are shown as a function of temperature in Figure 3.21b and c, respectively. The results showed that PTI nanopores can sieve He and H_2 from larger molecules. A high-temperature H_2 sieving from PTI nanosheet-based films is demonstrated, with H_2 permeance reaching 1,500 gas permeation units, with H_2/CO_2, H_2/N_2, and H_2/CH_4 selectivity reaching 10, 50, and 60, respectively, at 250°C.

3.5 CONCLUDING REMARKS

The photocatalytic activity of *g*-CN-related materials depends strongly on their precursor and their synthesis process. In Table 3.5, the H_2 evolution rates of *g*-CN-related materials are listed together with precursors, synthesis conditions, co-catalyst used, with or without sacrificial scavenger agent, and structural characteristics, including texture and morphology of the materials. Unfortunately, many of the studies are reporting the H_2-evolution rates on the basis of the amount of catalysts used, and so they had to be recalculated in the same unit "mmol/g·h". The values recalculated from the original values are indicated by using "[a]" marks.

Table 3.5 demonstrates that highly crystalline *g*-CN delivers a high H_2-evolution rate, higher than those reported on doped *g*-CNs. Using a single crystal of LiCl-intercalated CN, it has been reported that the prismatic $\{10\bar{1}0\}$ planes are photocatalytically more active than the basal {0001} planes [43]. In other words, the thickness of *g*-CN particles may be an important factor in discussing the photocatalytic activities, including H_2 evolution activity, of *g*-CN-related catalysts. On the photocatalytic activity of *g*-CN-related

TABLE 3.5
H_2-Evolution Rate of *g*-CN-Related Materials with Their Synthesis Conditions

Precursor	Synthesis Conditions	Texture and Morphology	H_2 Evolution: Co-Catalyst, Sacrificial Agent, etc.	H_2 Evolution: Rate (mmol/h·g)	References
		g-C_3N_4 and Modified g-C_3N_4			
Cyanamide	550°C, 4 h	Ground powder	Pt (0.5 wt%) water with TEA	0.073[a]	5
Dicyandiamide (DCDA)	Calcination at 550°C for 3 h		Pt (3 wt%) water without agent	0.0313	8
	Seventh re-calcination at 600°C for 2 h	Slight improvement in crystallinity		0.7648	
Urea	600°C	S_{BET} of 12.8–43.8 m^2/g	Pt (3 wt%) water with TEOA	19.412	9
DCDA				1.35	
Thiourea				2.47	
Urea	550°C, 3 h	S_{BET} of 11.3–69.6 m^2/g	Pt (3 wt%)	0.59[a]	10
DCDA				0.261[a]	
Thiourea				0.189[a]	
Melamine + cyanuric acid supramolecular aggregates	Microwave-assisted (16 min)	Highly crystalline	Pt (0.5 wt%) with TEOA	2.003[a]	11
Melamine	0.1 M HCl-washing	Highly crystalline	Pt (3 wt%) with TEOA	0.684	12
Melamine	KCl/LiCl, 550°C 4 h	Highly crystalline	Pt (3 wt%) with TEOA + K_2HPO_4	15.4[a]	13
Melamine	KCl/LiCl	Triazine/heptazine heterojunction	Pt (3 wt%) with TEOA	11.720	14

(*Continued*)

TABLE 3.5 (*Continued*)
H_2-Evolution Rate of *g*-CN-Related Materials with Their Synthesis Conditions

			H_2 Evolution		
Precursor	Synthesis Conditions	Texture and Morphology	Co-Catalyst, Sacrificial Agent, etc.	Rate (mmol/h·g)	References
Melamine	Mixed with KCl	Crystalline with K^+-intercalation	Pt (1 wt%) with TEOA	1.356	15
Melamine	$LiCl{\cdot}H_2O$–KCl–NaCl, 500°C	Intercalation compound of crystalline *nt*-CN	Pt (3 wt%), Salts-saturated water with 10 vol% TEOA	3.46[a]	16
HNO_3-treated melamine + cyanuric acid	550°C	Distorted cyanuric acid	Pt (1 wt%) with TEOA	7.46	19
Dicyandiamide + barbituric acid	550°C	N-rich (*C/N* of 0.74)	Pt (3 wt%) with TEOA	2.531[a]	20
Cyanamide + silica	550°C 4 h SiO_2-colloid templating	Mesoporous, S_{BET} of (67–373 m^2/g)	Pt (3 wt%) with TEOA	1.49[a]	22
g-CN with boric acid H_3BO_3	550°C	2.67 at% B-doped, tubular	Pt (3 wt%) with TEOA	4.4[a]	25
Self-assembled C-CN	Thiourea + trimesic acid	C-doped self-assembled supramolecular CN	Pt (3 wt%) in water with TEOA	1.405	26
Cyanuric + acid/melamine (1/1) + benzonitrile (100 μL)	550°C for 4 h	C and O co-doped	Pt (1 wt%)	3.77	27
Melem + formaldehyde	Copolymerization, 80°C	O-linked CN	With TEOA	1.745	6
Melem + phosphonitrilic chloride trimer	500°C, 4 h	P-linked CN	Without co-catalyst	0.388	28

(Continued)

TABLE 3.5 (*Continued*)
H_2-Evolution Rate of *g*-CN-Related Materials with Their Synthesis Conditions

			H_2 Evolution		
Precursor	Synthesis Conditions	Texture and Morphology	Co-Catalyst, Sacrificial Agent, etc.	Rate (mmol/h·g)	References
Urea + H_3PO_2 + $(NH_4)_2S_2O_8$	550°C, 2 h	O and P co-doped	Pt (3 wt%) in water with TEOA	0.44	29
Melamine + thiolactic acid ($C_3H_6O_2S$)	520°C 2 h in Ar	S and C co-doped	Pt (1 wt%) in water with TEOA	12.75	30
Ultrathin CN (*ut*-CN)	Solvothermal-exfoliation at 180°C	N-vacancy (V_{N2C})	Pt in water with TEOA	1.74	31
		***g*-C_3N_4-Based Composites**			
Dicyandiamide	500°C and 600°C, 4 h	Homojunction of CN-500/CN-600	Pt (2 wt%) in water with TEOA	0.3326[a]	32
Melamine-derived CN Ox-CN and Re-CN	Spray-calcination at 550°C, 2 h	Triple-layered Ox-CN/CN/Re-CN (Homojunctions)	Pt (6 wt%) in water with TEOA	127.05	33
S-CN-loaded onto $ZnIn_2S_4$	S-doped CN (S-CN)	$ZnIn_2S_4$-loaded S-doped CN	In water with 0.35 M Na_2SO_3 and 0.25 M Na_2S	8.463	36
A triazine-based COF (TFPT-DHTH-COF)		CdS-loaded COF	Pt (3 wt%) ascorbic acid (0.1 M)	4.57	37
g-CN + poly(3-hexylthiophene) (P3HT)	*g*-CN-impregnation into P3HT $CHCl_3$ solution	5 wt% P3HT-loaded/CN	Pt (1 wt%)	1.87[a]	39

[a] Recalculated to the common unit (mmol/g·h) by taking into account the amount of catalysts used.

materials, it has been believed that the heterojunctions, as well as homojunctions in *g*-CN materials, provide active sites. Therefore, numerous efforts and trials to create and control these junctions in *g*-CN materials have been reported, via various modifications of *g*-CN, such as doping, intercalation, exfoliation, and compositing with other materials. In these experiments, the photocatalytic effects were compared with the pristine *g*-CN in most cases and reported certain enhancements in photocatalytic activity so far. However, modification of highly crystalline *g*-CN has never been reported and has never been compared with the highly crystalline *g*-CN, as the present authors are concerned. Recently, the crystallinity of *g*-CN-related catalysts was pointed out to be an important factor to govern their photocatalytic activities by reviewing various published literature [44]. Most of the modifications of *g*-CN may introduce some distortion in the basic structure of *g*-CN, but some may work to create some heterojunctions. So, there might be a competition between the activity reduction due to the introduction of structural distortion and the activity increase due to the creation of heterojunctions.

In addition, it must also be pointed out that the experimental results on highly crystalline *g*-CNs are not many and the crystallinity of *g*-CNs has not been evaluated quantitatively. No definition of "crystallinity" for *g*-CNs and no quantitative evaluation method has been established. X-ray diffraction analysis is a commonly available technique for evaluating the crystallinity of *g*-CNs.

On carbon materials, graphite is the final and the highest crystallinity, which is evaluated as "graphitization degree" by the detailed analysis of two-dimensional 10 diffraction profiles [45]. For practical evaluation of a wide range of carbon materials, the interlayer spacing d_{002} is determined from diffraction angles of either the 002 or 004 diffraction profile. On the measurement of these X-ray diffraction profiles, a standard procedure was proposed [46] and has been widely employed in research laboratories and even in carbon industries. The specification proposed is strongly recommended to use a thin specimen and to employ an internal standard (practically a high-purity silicon powder) for X-ray diffraction measurements to avoid additional broadening and angle-shift of diffraction profiles due to low X-ray absorption coefficient and to increase reproducibility of the measurements in different laboratories and different researchers. By applying this procedure, the reliability and reproducibility of X-ray diffraction data are markedly strengthened, including d_{002} calculated from diffraction angles 2θ and crystallite sizes L_c and L_a calculated from half widths of respective diffraction profiles. In addition, for carbon materials, the magnetoresistance [47,48] and Raman spectrum analysis [49] have been proposed for crystallinity and nanotexture evaluation of a wide range of carbon materials. *g*-CNs are placed in the same situation as carbon materials, with low X-ray absorption and the easy occurrence of preferred orientation in a specimen for X-ray diffraction measurements.

REFERENCES

1. Ji Y, Yang M, Dong H, et al., Two-dimensional germanium monochalcogenide photocatalyst for water splitting under ultraviolet, visible to near-infrared light. *Nanoscale* 2017; 9: 8608–15.
2. Fu C-F, Li, X, Luo, Q, et al., Two-dimensional multilayer M_2CO_2 (M=Sc, Zr, Hf) as photocatalysts for hydrogen production from water splitting: A first principles study. *J Mater Chem A* 2017; 5: 24972–80.

3. Shan P, Geng K, Shen Y, et al., Facile synthesis of hierarchical core-shell carbon@ $ZnIn_2S_4$ composite for boosted photothermal-assisted photocatalytic H_2 production. *J Colloid Interface Sci* 2025; 677: 1098–107.
4. Rezaei M, Nezamzadeh-Ejhieh A, Massah AR, A comprehensive review on the boosted effects of anion vacancies in photocatalytic solar water splitting: Focus on g-C_3N_4 with carbon and nitrogen vacancies. *Energy Fuels* 2024; 38: 8406–36.
5. Maeda K, Wang XC, Nishihara Y, et al., Photocatalytic activities of graphitic carbon nitride powder for water reduction and oxidation under visible light. *J Phys Chem C* 2009; 113: 4940–7.
6. Yang S, Wang Q, Wang Q, et al., Linkage engineering mediated carriers transfer and surface reaction over carbon nitride for enhanced photocatalytic activity. *J Mater Chem A* 2021; 9(38): 21732–40.
7. Kumar P, Boukherroub R, Shankar K, Sunlight-driven water-splitting using two-dimensional carbon based semiconductors. *J Mat Chem A* 2018; 6: 12876.
8. Wu X, Cheng J, Li X, et al., Enhanced visible photocatalytic oxidation of NO by repeated calcination of g-C_3N_4. *Appl Surf Sci* 2019; 465: 1037–46.
9. Martin DJ, Qiu K, Shevlin SA, et al., Highly efficient photocatalytic H_2 evolution from water using visible light and structure-controlled graphitic carbon nitride. *Angew Chem Int Ed*, 2014; 53: 9240–5.
10. Zhang Y, Liu J, Wu G, et al., Porous graphitic carbon nitride synthesized via direct polymerization of urea for efficient sunlight-driven photocatalytic hydrogen production. *Nanoscale* 2012; 4: 5300–3.
11. Guo Y, Li J, Yuan Y, et al., A rapid microwave-assisted thermolysis route to highly crystalline carbon nitrides for efficient hydrogen generation. *Angew Chem Int Ed* 2016; 55: 14693–7.
12. Li Y, Zhang D, Feng X, et al., Enhanced photocatalytic hydrogen production activity of highly crystalline carbon nitride synthesized by hydrochloric acid treatment. *Chinese J Catal* 2020; 41: 21–30.
13. Lin L, Ou H, Zhang Y, Wang X, Tri-s-triazine-based crystalline graphitic carbon nitrides for highly efficient hydrogen evolution photocatalysis. *ACS Catal* 2016; 6: 3921–31.
14. Sun Z, Dong H, Yuan Q, et al., Self-supported hierarchical crystalline carbon nitride arrays with triazine-heptazine heterojunctions for highly efficient photoredox catalysis. *Chem Eng J* 2022; 435: 134865.
15. Xu Y, He X, Zhong, H, et al., Solid salt confinement effect: An effective strategy to fabricate high crystalline polymer carbon nitride for enhanced photocatalytic hydrogen evolution. *Appl Catal B* 2019; 246: 349–55.
16. Gao H, Yan S, Wang J, et al., Towards efficient solar hydrogen production by intercalated carbon nitride photocatalyst. *Phys Chem Chem Phys* 2013; 15: 18077–84.
17. Tang S, Wang W, Ansari HM, et al., A typical supramolecular self-assembly derived graphitic carbon nitride with $n \rightarrow \pi^*$ electron transition capable of efficient visible light hydrogen production. *Appl Surf Sci* 2024; 674: 160917.
18. Jun Y-S, Park J, Lee SU, et al., Three-dimensional macroscopic assemblies of low-dimensional carbon nitrides for enhanced hydrogen evolution. *Angew Chem Int Ed* 2013; 52: 11083–7.
19. Lei L, Fan H, Jia Y, et al., Cyanuric acid-assisted synthesis of hierarchical amorphous carbon nitride assembled by ultrathin oxygen-doped nanosheets for excellent photocatalytic hydrogen generation. *ACS Appl Mater Interfaces* 2024; 16(12): 14809–21.
20. Zhang J, Chen X, Takanabe K, et al., Synthesis of a carbon nitride structure for visible-light catalysis by copolymerization. *Angew Chem Int Ed* 2010; 49: 441–4.
21. Goettmann F, Fischer A, Antonietti M, et al., Chemical synthesis of mesoporous carbon nitrides using hard templates and their use as a metal-free catalyst for Friedel-Crafts reaction of benzene. *Angew Chem Int Ed* 2006; 45: 4467–71.

22. Wang X, Maeda K, Chen XF, et al., Polymer semiconductors for artificial photosynthesis: Hydrogen evolution by mesoporous graphitic carbon nitride with visible light. *J Am Chem Soc* 2009; 131: 1680–1.
23. Kailasam K, Epping JD, Thomas A, et al., Mesoporous carbon nitride–silica composites by a combined sol-gel/thermal condensation approach and their application as photocatalysts. *Energy Environ Sci* 2011; 4: 4668–74.
24. Chen X, Jun Y-S, Takanabe K, et al., Ordered mesoporous SBA-15 type graphitic carbon nitride: A semiconductor host structure for photocatalytic hydrogen evolution with visible light. *Chem Mater* 2009; 21: 4093–5.
25. Zhang S, An C, Zhang R, et al., Template-free synthesis of boron-doped graphitic carbon nitride porous nanotubes for enhanced photocatalytic hydrogen evolution. *Langmuir* 2024; 40(12): 6453–62.
26. Dharmarajan NP, Fawaz M, Sathish C, et al., Insights into atomic level π-electron modulations in supramolecular carbon nitride nanoarchitectonics for sustainable green hydrogen production. *Adv Energy Mater* 2024; 14(29): 2400686.
27. Liu D, Zhao C, Chen M, et al., Enhanced visible light photocatalytic performance of carbon and oxygen co-doped carbon nitride with a three-dimensional structure: Performance and mechanism study. *J Colloid Interfaces Sci* 2024; 665: 452–64.
28. Song M, Song H, Chen P, et al., Polarization reorientation of carbon nitride induced by a local bridge donor for boosting photocatalytic H_2 evolution integrated with selective amine oxidation. *ACS Sustain Chem Eng* 2023; 11: 8096–195.
29. Hoang TVA, Nguyen PA, Pham TT, et al., Homogeneous distribution of highly dispersed Pt species over O and P co-doped g-C_3N_4 and its superior photocatalytic H_2 evolution activity. *J Alloys Compd* 2024; 997: 174923.
30. Zhao B, Gong W, Liu X, et al., Ternary Synergism: Synthesis of S, C Co-doped g-C_3N_4 hexagonal ultra-thin tubular composite photocatalyst for efficient visible-light-driven photocatalytic hydrogen production. *Inter J Hydrogen Energy* 2024; 61: 1317–29.
31. Pang Y, Zhang R, Li L, et al., Anchoring Pt single atoms on specific nitrogen vacancies of carbon nitride to accelerate photogenerated carrier transfer. *J Colloid Interface Sci A* 2025; 677: 1–10.
32. Khan MA, Rameel MI, Salam F, et al., Construction of isotype heterojunctions in polymeric carbon nitride with thermal modulation and improved photocatalytic hydrogen production activity. *Mater Chem Phys* 2024; 315: 129052.
33. Zhou Q, Guo Y, Ye Z, et al., Carbon nitride photocatalyst with internal electric field induced photogenerated carriers spatial enrichment for enhanced photocatalytic water splitting. *Mater Today* 2022; 58: 100–9.
34. Li H, Liu Y, Gao X, et al., Facile synthesis and enhanced visible-light photocatalysis of graphitic carbon nitride composite semiconductors. *ChemSusChem* 2015; 8: 1189–96.
35. Li Q, Yue B, Iwai H, et al., Carbon nitride polymers sensitized with N-doped tantalic acid for visible light-induced photocatalytic hydrogen evolution. *J Phys Chem C* 2010; 114: 4100–5.
36. Wang T, Pan X, He M, et al., In situ construction of hollow coral-like porous S-doped g-C_3N_4/$ZnIn_2S_4$ S-scheme heterojunction for efficient photocatalytic hydrogen evolution. *Adv Sci* 2024; 11: 2403771.
37. Zhang X, Wu X, Chen R, et al., A triazine-based covalent organic framework decorated with cadmium sulfide for efficient photocatalytic hydrogen evolution from water. *J Colloid Interface Sci* 2024; 665: 100–8.
38. Zhu W, Qin L, Ou Z, et al., Protonated carbon nitride cooperated with iron porphyrin metal-organic framework for boosting photocatalytic hydrogen production and degradation of tetracycline. *Fuel* 2024; 376: 132665.
39. Yan H, Huang Y, Polymer composites of carbon nitride and poly(3-hexylthiophene) to achieve enhanced hydrogen production from water under visible light. *Chem Commun* 2011; 47: 4168–70.

40. Chu S, Wang C, Feng J, et al., Melem: A metal-free unit for photocatalytic hydrogen evolution. *Int J Hydrogen Energy* 2014; 39: 13519–26.
41. Qi S, Fa, Y, Wang J, et al., Metal-free highly efficient photocatalysts for overall water splitting: C_3N_5 multilayers. *Nanoscale* 2020; 12: 306–15.
42. Villalobos LF, Vahdat MT, Dakhchoune M, et al., Large-scale synthesis of crystalline g-C_3N_4 nanosheets and high-temperature H_2 sieving from assembled films. *Sci Adv* 2020; 6(4): 9851.
43. Lin L, Lin Z, Zhang J, et al., Molecular-level insights on the reactive facet of carbon nitride single crystals photocatalysing overall water splitting. *Nat Catal* 2020; 3: 649–55.
44. Li H, Cheng B, Xu J, et al., Crystalline carbon nitrides for photocatalysis. *EES Catal* 2024; 2: 411–47.
45. Warren BE, X-ray diffraction in random layer lattices. *Phys Rev* 1941; 9: 693–8.
46. Iwashita N, Park CR, Fujimoto H, et al., Specification for a standard procedure of X-ray diffraction measurements on carbon materials. *Carbon* 2004; 42: 701–14.
47. Hishiyama Y, Kaburagi Y, Inagaki M, Characterization of structure and microtexture of carbon materials by magnetoresistance technique. In *Chemistry and Physics of Carbon*, Thrower PA Ed., Macel Dekker, New York, 1971; Vol. 23: pp. 1–68.
48. Hishiyama Y, Kaburagi Y, Inagaki M, Magnetoresistance. In *Materials Science and Engineering of Carbon: Characterization*, Inagaki M and Kang Y Eds., Tsinghua University Press and Elsevier, Beijing, China and Amsterdam, Netherlands, 2017; pp. 173–204.
49. Kaburagi Y, Yoshida A, Hishiyama Y, Raman spectroscopy. In *Materials Science and Engineering of Carbon: Characterization*, Inagaki M and Kang Y Eds., Tsinghua University Press and Elsevier, Beijing, China and Amsterdam, Netherlands, 2017; pp. 125–52.

4 Carbon Dioxide Conversion to Hydrocarbons

For the reduction of CO_2 to C1/C2 hydrocarbon fuels, a variety of heterogeneous photocatalysts based on inorganic compounds have been proposed, including Pd-loaded TiO_2 [1], Ag-loaded $ALa_4Ti_4O_{15}$ (A=Ca, Sr, and Ba) [2], layered double hydroxides [3], Cu single-atom-loaded through N-doped carbon onto TiO_2 [4], WO_3/BiOBr [5], and others. Many review articles have also been published on this topic [6,7]. The collective thermo-catalytic/photocatalytic/electrocatalytic conversion of CO_2 to chemicals and fuels using *g*-CN was reviewed [8,9].

In this chapter, the recent results on photocatalytic conversion to C1/C2 hydrocarbon fuels on *g*-C_3N_4-based photocatalysts, including the modified and composited materials, are summarized.

4.1 *g*-C_3N_4 AND MODIFIED *g*-C_3N_4

Acetonitrile solution of a mixture of guanidine hydrochloride (CH_6ClN_3, GH) with dicyandiamide (DCDA) (1/1 by mol) was heated under solvothermal conditions at 180°C for 48 h to synthesize CN nanosheets rich in cyano- and carboxyl-functions [10], which was coded as C-CN. For comparison, bulk CN (*b*-CN) was prepared from melamine by calcination at 550°C for 2 h. Photocatalytic CO_2-reduction performances of two photocatalysts on 100 mg were measured in a sealed reactor containing CO_2 generated by the reaction of H_2SO_4 with $NaHCO_3$ at a humidity of 41% under the irradiation of light from a 350 W Xe lamp. In Figure 4.1a, the CO_2 adsorption isotherms of C-CN and *b*-CN are shown. The C-CN can adsorb a much larger amount of CO_2 and the CO_2 temperature-programmed desorption measurement demonstrated that the CO_2 molecules were more tightly adsorbed on the C-CN surface. In Figure 4.1b, the production rates of the gases, CO, CH_4, CH_3OH, and CH_3CH_2OH, produced by photocatalytic reduction of CO_2 over two catalysts are compared. CO-formation was much suppressed over C-CN, while the conversion to hydrocarbon gases, i.e., CH_4, CH_3OH, and CH_3CH_2OH, was much accelerated on the C-CN. The yield of H_2 over C-CN and *b*-CN was measured as only 0.39 and 0.12 μmol/h·g, respectively, demonstrating the high selectivity of CO_2 photoreduction over these catalysts.

Pore structure of urea-derived *g*-CN powder was modified by hydrothermal treatment in glycerol/water mixed solvent (0/60, 10/50, 30/30, 50/10, and 60/0 by volume) at 160°C for 10 h [11]. With increasing the amount of glycerol *X* (0–60), pore structure parameter increases up to *X*=50 (coded as CN-50) and the light absorbance spectrum in 420–800 nm region exhibited a slight red-shift owing to

DOI: 10.1201/9781003589051-4

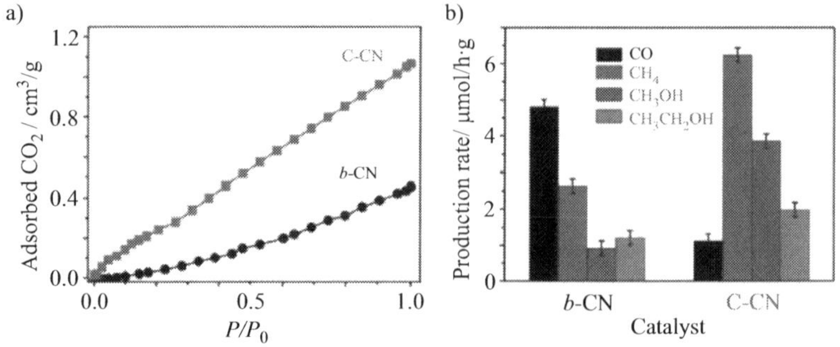

FIGURE 4.1 CO_2-conversion performances of *b*-CN and C-CN: (a) CO_2 adsorption isotherm and (b) the production rate of CO, CH_4, CH_3OH, and CH_3CH_2OH [10].

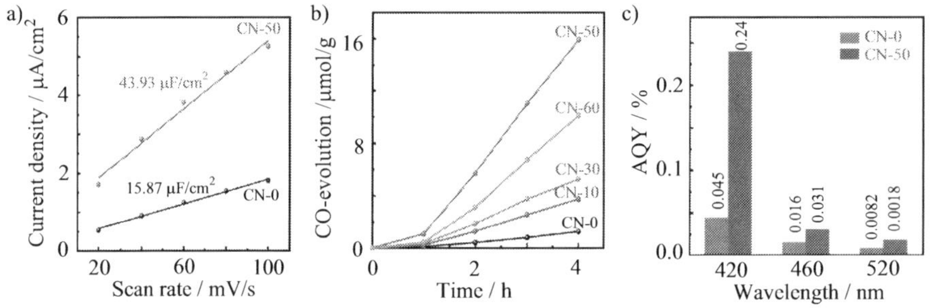

FIGURE 4.2 *g*-CNs with modified pore structures: (a) current density vs. scan rate on double-layer capacitors, (b) CO evolution kinetics under visible light irradiation, and (c) apparent quantum yield (AQY) at different wavelengths of light [11].

the introduction of three C-coordinated N-vacancies (V_{N3C}). In order to evaluate whether glycerol treatment creates new active sites for CO_2 reduction, the electrochemical active surface area (ECSA) and the density of catalytic active sites were determined by the measurement of double-layer capacitance. In Figure 4.2a, the current density of the capacitor against scan rate of CV measurements is shown on CN-0 (the pristine CN) and CN-50 (glycerol-treated CN), revealing that the latter gave significantly higher capacitance (43.93 μF/cm^2) than the former did (15.87 μF/cm^2). This result suggests that the glycerol treatment provides more ECSA by the introduction of V_{N3C} vacancies in the *g*-CN. Photocatalytic CO_2-reduction reaction was studied as a gas/solid reaction without any sacrificial agent, and visible light ($\lambda > 420$ nm) was used as the light source. The CO-evolution curves are shown on the resultant glycerol-modified CNs (CN-10 to CN-60) in comparison with the pristine CN (CN-0), as shown in Figure 4.2b. The CN-50 catalyst demonstrates the highest CO-evolution. The CO-evolution rate reached 4.18 μmol/h·g, about 12 times larger than CN-0 (0.32 μmol/h·g). The apparent quantum yield (AQY) was measured under monochromatic light irradiation (Figure 4.2c). The AQY of the CN-50 is calculated to be 0.24% under 420 nm light irradiation, while the AQY is only 0.045% for CN-0. Under 460 and 520 nm light irradiation, the AQY of 50-CN is also higher than that of

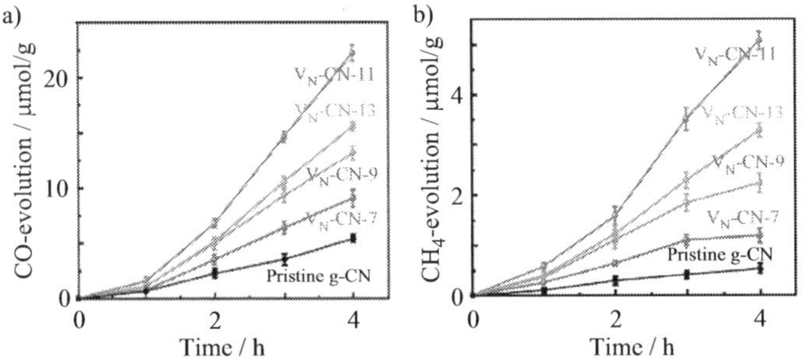

FIGURE 4.3 CO_2-conversion performances of *g*-CNs with abundant N-vacancies (V_N-CNs); the conversion kinetics to (a) CO and (b) CH_4 [13].

0-CN. It was concluded that the glycerol treatment of *g*-CN resulted in an excellent photocatalytic CO_2 reduction activity, which may be caused by the rich V_{N3C} on the surface of the catalysts. The CN-50 catalyst was subjected to cycling tests to assess its reusability.

The creation of V_{N3C} in the center of the heptazine rings of *g*-CN via a photochemically assisted N-etching was shown to be effective for enhancement in CO_2 conversion [12]. The photochemical etching was performed using different light radiation times (1, 3, 5, and 10 h) to produce different N-vacancy (V_N) concentrations and the resultant N-deficient *g*-CN were coded as V_N-CN-*X*, where *X* corresponds to light irradiation time. The V_N-CN-5 delivered the highest CO-generation, achieving a CO-generation rate of 35.2 µmol/g·h, being 3.4 times higher than that of the pristine *g*-CN (10.4 µmol/g·h), with nearly 100% selectivity.

A series of *g*-CN with abundant N-vacancies were prepared using a green and facile strategy and studied for the CO_2 conversion and reduction of Cr^{4+} [13]. The pristine *g*-CN was synthesized from melamine at 550°C for 3 h, and it (0.5 g) was mixed with various amounts of sodium bisulfite $NaHSO_3$ (7, 9, 11, and 13 g) in 50 mL of deionized water under stirring for 24 h and filtering, washing many times until no Na^+ was detected, the resultant V_N-CN being coded as V_N-CN-7, -9, -11, and -13. The chemical composition determined by EDS, $S_{BET,}$ and bandgap changed from the pristine *g*-CN to V_N-CN-11, a marked increase in atomic ratio *C*/*N* from 0.90 to 1.29, while a slight increase in S_{BET} from 35 to 49 m^2/g and bandgap from 2.68 to 2.77 eV. In Figure 4.3a and b, the performances of CO_2 conversion to CO and CH_4 are compared on the resultant series of V_N-C_N-*X* photocatalysts. Both CO and CH_4-evolution performances are enhanced by the introduction of N-vacancies, depending strongly on the amount of $NaHSO_3$ used, i.e., the content of N-vacancies in the *g*-CN structure, suggesting the optimum content must be selected for a noticeable acceleration of CO_2 conversion. The catalyst V_N-CN-11 exhibited a high reduction kinetics of Cr^{4+}, much faster than the pristine CN, revealing a marked acceleration in Cr^{4+}-reduction with N-vacancy introduction.

B-doped *g*-CN was synthesized by calcining a mixture of *g*-CN, which was prepared from melamine at 550°C for 4 h, with different amounts of BH_3NH_3 at 600°C in N_2 flow and coded as B_X-CN, where *X* is at% of B-doped (0.5–3.0 at%) [14].

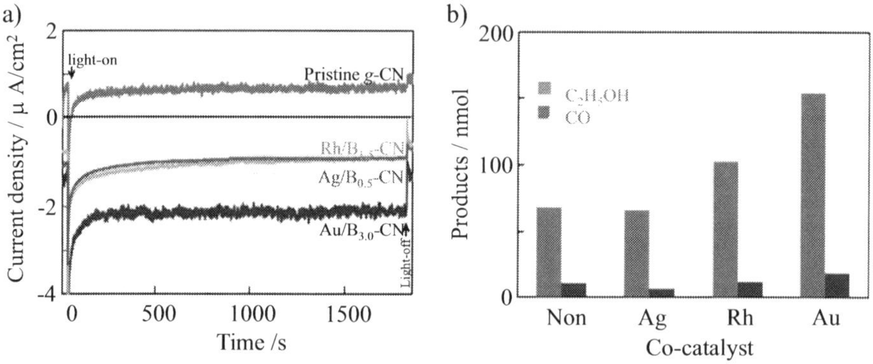

FIGURE 4.4 Photelectrochemical CO_2 conversion over the B-doped CNs (B_X-CNs) in comparison with the pristine *g*-CN: (a) time dependence of a cathodic photocurrent (chronoamperometry) and (b) products over the $B_{3.0}$-CN catalyst with different co-catalyst-loading [14].

Thin film electrodes of the pristine CN and B_X-CNs were deposited on the surface of FTO substrates by electrophoresis, followed by loading a co-catalyst (Au, Ag, Cu, or Rh) with magnetron sputtering. Their photocatalytic performances for CO_2 reduction were measured in a three-electrode cell using the CN-based electrode (the cathode), a glassy carbon electrode, and an Ag/AgCl electrode in a 0.5 M $NaHCO_3$ solution (pH 7.3) with CO_2 bubbling under a solar-simulated light for 120 min. As shown in Figure 4.4a, the cathodic photocurrent increased suddenly at the beginning, becoming relatively stable after 30 min. The $B_{3.0}$-CN catalyst exhibited the highest photocurrent, suggesting that the optimum B-doping is 3.0 at%, with a cathodic photocurrent density of 6.0 µA/cm² at an applied potential of −0.40 V *vs.* Ag/AgCl. The products of photoelectrochemical reduction of CO_2 over the $B_{3.0}$-CNs loaded with different co-catalysts were measured by chronoamperometry. The results depend on the co-catalyst used, as shown in Figure 4.4b, revealing that the principal product is C_2H_5OH and its yield increases markedly with co-catalyst loaded. The co-catalyst Au delivered the highest Faraday efficiencies of 47 and 0.9% for C_2H_5OH and CO, respectively. The production of C_2H_5OH from CO_2 gas over the $B_{3.0}$-CN electrode was confirmed using the labeled $^{13}CO_2$ gas by ^1H NMR spectroscopy.

P- and C-co-doped CN (P_XC-CN) was synthesized in a two-step process [15]; first, the aqueous solution of 6 g DCDA and 0.2–0.6 g phytic acid ($C_6H_{18}O_{24}P_6$, PA, 70% concentration) at 85°C was hydrothermally treated at 180°C for 10 h to synthesize the supramolecular precursors and then calcined at 550°C for 2 h to obtain P_XC-CN, where *X* corresponds to the amount of PA used (0.2–0.6 g). Estimated positions of P and excess C in *g*-CN are illustrated in Figure 4.5a as an inset. The CO_2-conversion performances were measured under the irradiation of the visible light from a 300 W Xe-lamp equipped with a 420 nm cutoff filter using 25 mg catalyst powder dispersion in water containing 10 vol% triethanolamine (TEOA) with CO_2 (≥99.99%) bubbling, and CO_2-conversion product was predominantly CH_4, accompanied by traces of CO and H_2. CH_4-generation kinetics over the resultant P_XC-CN catalysts are shown in Figure 4.5a, revealing $P_{0.4}$C-CN delivered the highest CH_4 generation, of which the rate reached 41.84 µmol/g·h, about 12 times higher than that of the pristine CN (3.43 µmol/g·h) after 5 h irradiation, as shown in Figure 4.5b.

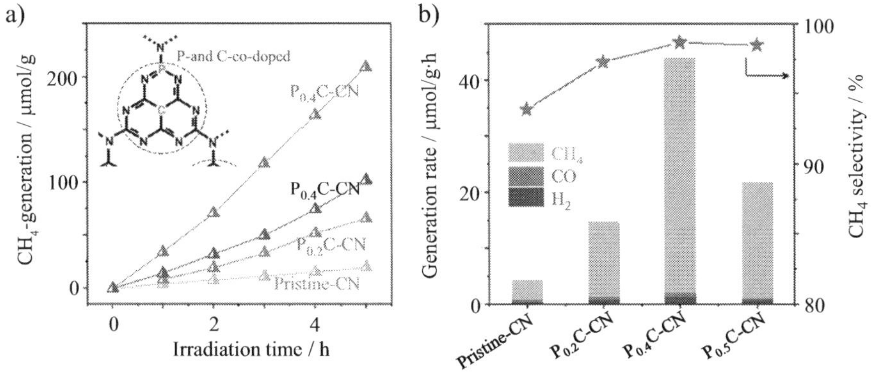

FIGURE 4.5 CO_2-conversion performances of P- and C-co-doped CN (P_XC-CN): (a) CH_4-generation kinetics with illustration of P- and C-doping positions (inset) and (b) the average generation rates of the reaction products, CH_4, CO, and H_2 after 5 h irradiation and CH_4 selectivity [15].

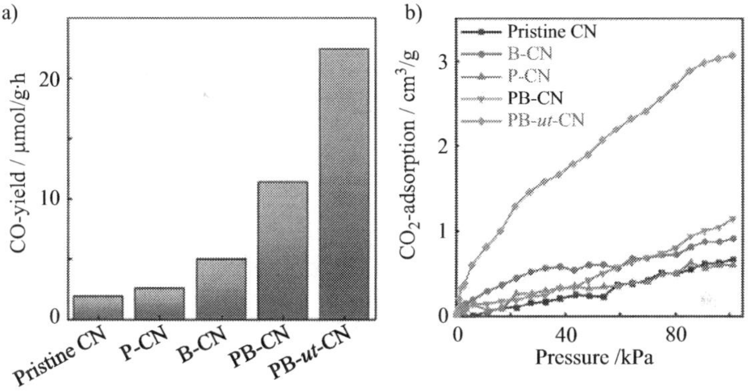

FIGURE 4.6 P- and B-co-doped ultrathin CN (PB-*ut*-CN): (a) CO-yield over the related catalysts and (b) CO_2-adsorption behaviors of the catalysts [16].

P- and B-co-doping was shown to be efficient for improving the CO_2-conversion performance of *g*-CN, and the thinning of *g*-CN resulted in further enhancement in the CO_2-reduction reaction [16]. A mixture of melamine (1 g), NH_4Cl (1 g), boric acid (0.04 g), and 2-aminoethylphosphonic acid (2-AEPA, 0.016g) was heated at 500°C for 3 h in N_2 flow and then further heated at 550°C for 5 h. 2-AEPA works as a phosphorus source as well as an in situ porogen promoting the formation of macroporous networks, and in addition, it releases NH_3 gas, which can modify the morphology of CN by inducing delamination (exfoliation) of stacked layers to an ultrathin sheet of CN (*ut*-CN) under the assistance of HCl and NH_3 gases formed from NH_4Cl decomposition. NH_4Cl plays a leading role in the introduction of surface amino functional groups and in the creation of mesopores as a second porogen during the polymerization of melamine. The CO-yield over the resultant catalysts under visible light irradiation and CO_2 adsorption capacity are shown in Figure 4.6a and b, respectively.

As shown in Figure 4.6a, CO-yield enhanced by co-doping of P and B, CO yield of 1.89 μmol/g·h for the pristine CN was enhanced to 2.54 and 4.94 μmol/g·h on P-CN and B-CN, respectively, and further enhanced up to 11.36 μmol/g·h on P- and B-co-doping, i.e., PB-CN. In the catalyst exfoliation process, the PB-*ut*-CN catalyst exhibited further enhancement in CO-yield to 22.45 μmol/g·h. The CO_2-adsorption capacity of the catalysts increased markedly by co-doping, as shown in Figure 4.6b.

g-CN having dual active sites, i.e., Cu single atom (Cu1) and N-vacancies V_{N2C}, was synthesized [17]. Thiourea (10 g) was dissolved in 30 mL deionized water, followed by a small amount of $CuCl_2$ addition and stirring for 3 h. The resulting white crystals were collected and heated at 550°C for 3 h to obtain $(Cu1)_X$-V_{N2C}-CN, where *X* corresponds to (Cu1) mass percent to CN (0.1–0.7). It could achieve a high CO-yield of 11.12 μmol/g·h with CO-selectivity of 98.50% without any sacrificial agent addition in water. The single Co atom and single Mn atom co-doped CN ((Co1)(Mn1)-CN) was synthesized via a supramolecular self-assembly formed from melamine and trithiocyanuric acid had dual single-atom active sites for CO_2 reduction to CO and H_2O oxidation to O_2 [18]. The composite (Co1)(Mn1)-CN exhibited the CO-generation rate of 47 μmol/g·h, much higher than the related CNs, i.e., (Co1)-CN. (Mn1)-CN, including the pristine CN, revealing that the co-doping was effective in enhancing CO_2 reduction. The coordination structure of single Cu-atom (Cu1) and P-doped polymeric CN (P-CN) were regulated by modulation the sequence of phosphidization and Cu deposition into CN, resulting the different CNs containing the (Cu1) coordinated by 3 N coded as $(Cu1)_{3N}$ and coordinated 3 P $(Cu1)_{3P}$, i.e., $(Cu1)_{3N}$-CN, $(Cu1)_{3N}$-P-CN, $(Cu1)_{3P}$-CN and $(Cu1)_{3P}$-P-CN [19]. Positions of (Cu1) in these composites are illustrated by highlighting (Cu1) in Figure 4.7a. The composite $(Cu1)_{3N}$-P-CN exhibited high CO_2-conversion performances, as shown in Figure 4.7a and b. The CO_2-reduction product was only CO, with no carbon-based products being detected. The pristine CN and P-CN showed a low CO production (about 1.0 μmol/g·h). An introduction of (Cu1) in P-CN in the form of $(Cu1)_{3N}$, namely, $(Cu1)_{3N}$-P-CN, sharply promoted the CO-production rate to about 18.5 μmol/g·h, as compared to $(Cu1)_{3N}$-CN (about 5.5 μmol/g·h). When (Cu1) was presented in the form of $(Cu1)_{3P}$ in $(Cu1)_{3P}$-P-CN, the CO-production rate

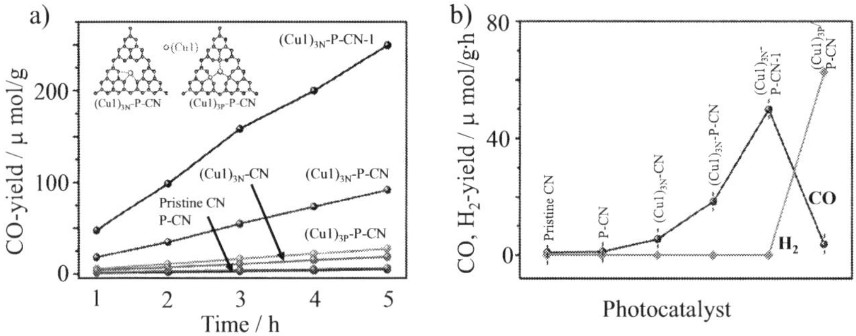

FIGURE 4.7 CO and H_2-evolution performances of (Cu1)- and P-co-doped CNs ($(Cu1)_{3N}$-CN, $(Cu1)_{3N}$-P-CN, $(Cu1)_{3P}$-P-CN in and $(Cu1)_{3N}$-P-CN-1) comparison with the pristine CN and P-CN: (a) CO-yield kinetics (1 atm, 20°C) with structure illustrations of the composites (insets) and (b) CO- and H_2-yields of the catalysts [19].

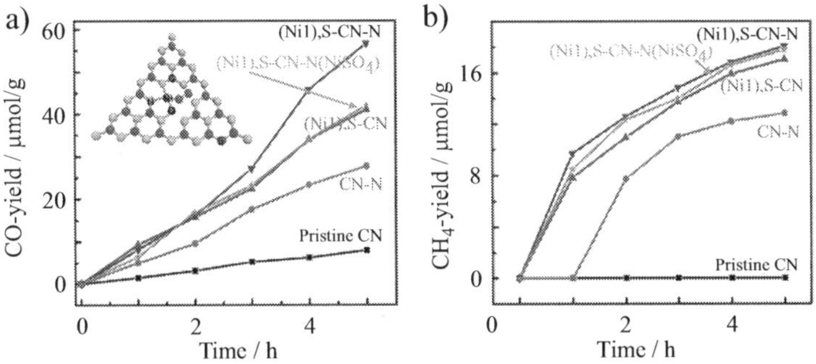

FIGURE 4.8 CO_2-conversion performances of the single Ni atom and S-co-doped *g*-CN, (Ni1), S-CN in comparison with the related catalysts: (a) CO-generation kinetics in comparison with the related composite and the pristine CN and (b) CH_4-generation kinetics [20].

(about 3.7 μmol/g·h) was greatly inhibited. However, in contrast, H_2 evolution over $(Cu1)_{3P}$-P-CN composite was greatly enhanced (62.5 μmol/g·h) from water splitting.

Single Ni atom (Ni1) and S-co-doped *g*-CN was synthesized by calcining the Ni-thiourea complex Ni(thiourea)$_6$$(NO_3)_2$ (Ni-TU) at 520°C [20]. Ni-TU was prepared by mixing $Ni(NO_3)_2{\cdot}6H_2O$ (14.54 g) with thiourea (15.22 g) in 25 mL of isopropyl alcohol at 90–100°C (near boiling point). An acetone solution of Ni-TU mixed with an additional urea, of which the precipitates were calcined at 520°C for 2.5 h to form the composite (Ni1), S-CN. Ni content in the composite was adjusted to 3 wt% of CN. The resultant composite (Ni1), S-CN of 100 mg and 500 mg NH_4Cl was mixed in 10 mL of water at 80°C and then freeze-dried, followed by calcination at 400°C for 2 h, which was coded as (Ni1), S-CN-N. The molecular structure is illustrated in Figure 4.8a as an inset. For comparison, the pristine CN and CN-N were prepared by the same procedure without the addition of Ni-TU. The resultant composites (10 mg) were coated onto the surface of FTO glass by making its slurry using deionized water. The catalyst-coated glass was placed in a reaction chamber with 0.5 mL of water and contacted with CO_2 gas at 8 kPa pressure under the irradiation of a simulated solar light from a 300 W Xe-lamp. Over the resultant composite, the products of CO_2 reduction were mainly CO and CH_4. In Figure 4.8a, CO-generation kinetics are compared with the related composites, including the pristine CN. The highest CO yield was obtained over the composite (Ni1)-S-CN-N as 58.6 μmol/g, much higher than that over the CN-N (27.7 μmol/g). The highest CH_4 is also given by (Ni1)-S-CN-N as 18.9 μmol/g, larger than that over the CN-N (12.8 μmol/g), as shown in Figure 4.8b. The pristine CN exhibited CO- and CH_4-yield of 7.9 and 0 μmol/g. These results demonstrated the effectiveness of the co-doping of a single Ni atom with S. Single rare-earth metal atom (practically Er-single atom)-doping was also effective to enhance CO_2-conversion performances [21]. A melamine sponge was immersed in an aqueous solution containing urea and $Er(NO_3)_3{\cdot}5H_2O$ under vigorous stirring to ensure uniform saturation with the precursors. The saturated sponge was then rapidly frozen using liquid nitrogen under dark conditions, followed by freeze-drying and then annealing in N_2 atmosphere. This process yielded a single Er atom (Er1) anchored in the *g*-CN nanotube (*nt*-CN), coded as $(Er1)_{6N}$/*nt*-CN. The Er atoms are reasonably assumed to be accommodated in the intrinsic cavity V_{6N}. Two composite materials with different (Er1) densities were prepared, having Er contents of 2.5 and 20.1 wt%, respectively. CO_2-conversion performances of these two composites were measured under simulated solar light irradiation in pure water. On the composite with the low-density (Er1), a CO-generation rate of 12.75 μmol/g·h and a CH_4-generation rate of 0.45 μmol/g·h, while the composite with the high-density (Er1) delivered 47.1 and 2.5 μmol/g·h, respectively.

P-doped CN (P-CN) having single Co atom and single Ag atom at adjacent sites (coded as (Co1)(Ag_{1+n})-P-CN) was synthesized [22], and its CO_2-reduction performances were compared with the composites with single Ag atom composite ((Ag1)-CN) and the composite having (Co1) and (Ag1) at dual metal sites ((Co1)(Ag1)-P-CN). The composite, (Co1)(Ag_{1+n})-P-CN, was prepared by solvothermal treatment of (Ag1)-CN in the presence of $NaH_2PO_2{\cdot}H_2O$ and $CoCl_2$. The optimized (Co1)(Ag_{1+n})-P-CN composite achieved a high CO-generation of 46.82 μmol/g with 70.1% selectivity in solid–liquid mode without any sacrificial agents after 4 h irradiation.

Single Pt and Cu atoms were doped in crystalline *g*-CN [23]. The crystalline *g*-CN (*cr*-CN) was synthesized from DCDA-derived *g*-CN using a molten salt KCl-LiCl at 550°C for 4 h. Single Pt and Cu atoms co-doped *cr*-CN ((Pt1), (Cu1)-*cr*-CN) were synthesized by first dispersing *cr*-CN (0.5 g) in deionized water with H_2PtCl_6 (50 mg) under stirring for 6 h. Subsequently, $CuCl_2$ (20 mg) was added, and the mixture was stirred for an additional 6 h. The resulting suspension was then frozen at approximately −10°C, followed by photoreduction using a 300 W Xe lamp for 10 min. On the resultant composites including pristine *cr*-CN, CO- and CH_4-generation kinetics are shown in Figure 4.9a and b, respectively, which were determined by 25 mg catalysts spreading on a Petri dish via ethanol solution slurry and sealing in reactor chamber and then the chamber was filled with CO_2 (99.99%, 70–80 kPa) and irradiated by a simulated solar light from a 300 W Xe lamp. The composite (Pt1), (Cu1)-*cr*-CN gave the highest performance for both CO- and CH_4-generations, higher than (Cu1)-*cr*-CN, (Pt1)-*cr*-CN, and *cr*-CN, revealing the effectiveness of co-loading for the enhancement of the CO_2 reduction. For (Pt1)(Cu1)-*cr*-CN composite, CO- and CH_4-yields after 3.5 h irradiation

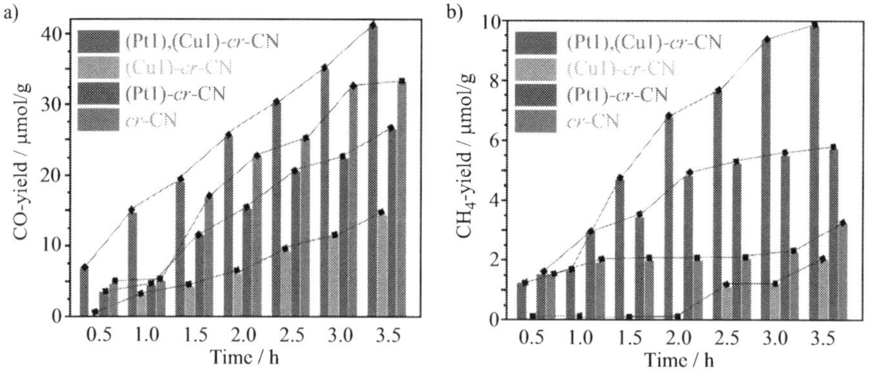

FIGURE 4.9 CO_2-reduction performances of single Cu and Pt atoms co-doped crystalline *cr*-CN composites (Pt1),(Cu1)-cr-CN in comparison with the related composites: (a) CO-generation kinetics and (b) CH_4-generation kinetics [23].

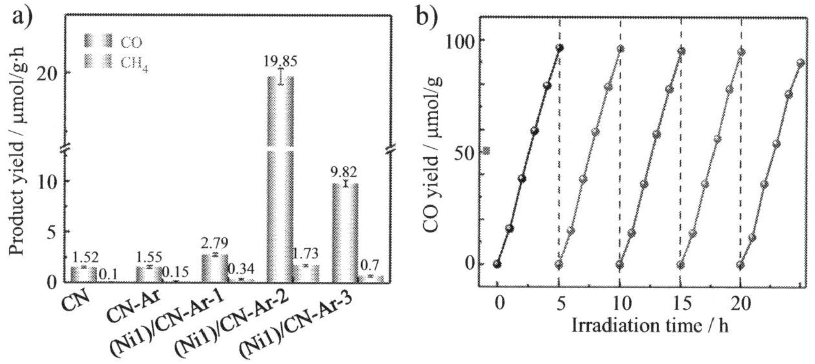

FIGURE 4.10 Single atom Ni-loaded *g*-CN catalysts, (Ni1)-CN-Ar-X, prepared via metal-vapor exfoliation procedure: (a) production yields of CO and CH_4 through CO_2 reduction by different catalysts and (b) cycle stability of CO production on Ni/CN-Ar-2 catalyst [24].

reached 41.1 and 9.8 μmol/g, respectively. Pt-containing composites ((Pt1), (Cu1)-*cr*-CN, and (Pt1)-*cr*-CN) exhibited relatively high selectivity toward CH_4.

The single Ni-atom loaded *g*-CN catalysts were prepared through Ni-vapor exfoliation at 550°C for different times in Ar (coded as (Ni1)-CN-Ar-X, where X is the holding time at 550°C) [24]. Their photocatalytic CO_2-conversion performances were studied by suspending 20 mg of the catalyst in 10 mL of deionized water and introducing high-purity CO_2 (99.999%) into the system under the irradiation with a 300 W Xe lamp from the top of the reactor for 4 h. As shown in Figure 4.10a, the main product of photocatalytic CO_2 reduction using CN-based catalysts was CO, associated with a small amount of CH_4, revealing that the exfoliation of catalyst, (Ni1)-CN-Ar-X, exhibited much higher activity than the pristine CN, probably owing to a larger specific surface area and the effect of single Ni particle doping, particularly (Ni1)-CN-Ar-2 delivering the production rates of 19.85 μmol/g·h for CO and 1.73 μmol/g·h for CH_4. Stability of CO production over the (Ni1)-CN-Ar-2 catalyst is shown as a cycle performance in Figure 4.10b, the rate of CO production being kept almost constant after 5 cycles of reaction.

Ultrathin *g*-CN-nanosheets (*ut*-CN) prepared by exfoliation of urea-derived CN in nitric acid, which were converted to boron-oxo species-doped CN (OB-*ut*-CN) by hydrothermal treatment in boric acid solution. Onto the resultant OB-*ut*-CN nanosheets, a single Ni atom was doped from $NiNO_3$ solution to form (Ni1)-OB-*ut*-CN composite [25]. The optimized (Ni1)-doped CN, $(Ni1)_{0.7}$-OB_5-*ut*-CN, provided excellent CO_2-reduction rate with CO and CH_4 as products, the CO- and CH_4-generation rates being calculated to be 22.1 and 8.7 μmol/g·h, respectively.

4.2 *g*-C_3N_4-BASED COMPOSITES

The composite CeO_2-loaded *g*-CN was synthesized via a hard templating with 40% SiO_2 colloid (average particle size of 12 nm) by calcination of a mixture of urea with *X* wt% $Ce(NO_3)_3·6H_2O$ (X = 1–5) at 550°C for 2 h followed by washing with NH_4HF_2

TABLE 4.1
Bandgap, Pore Structure Parameters, and Yields of CO and CH_4 for $(CeO_2)_X$/CN Composites in Comparison with Mesoporous CeO_2 [26]

Catalyst	Bandgap (eV)	S_{BET} (m^2/g)	Pore Size (nm)	CO Yield[a] (μmol/h)	CH_4 Yield[a] (μmol/h)
Pristine CN	2.72	121.0	10.4	0.339	0.010
$(CeO_2)_1$/CN	2.70	141.3	8.4	0.439	0.339
$(CeO_2)_2$/CN	2.68	147.8	9.3	0.590	0.454
$(CeO_2)_3$/CN	2.65	179.1	10.6	0.508	0.694
$(CeO_2)_4$/CN	2.67	155.8	9.1	0.386	0.226
$(CeO_2)_5$/CN	2.76	153.4	8.8	0.493	0.139
Mesoporous CeO_2	2.77	63.9	10.2	0.068	0.081

[a] After 1 h irradiation.

to remove SiO_2 to prepare the mesoporous composites (coded as $(CeO_2)_X$/CN) [26]. 50 mg photocatalyst powder was uniformly dispersed on a flat glass plate with a 1.5 cm × 1.5 cm area, which was exposed to a mixture of high-purity CO_2 and water vapor at 25°C under a light from a 300 W Xe-lamp. In Table 4.1, pore structure parameters and yields of CO and CH_4 after 1 h irradiation are listed. The composite $(CeO_2)_2$/CN and $(CeO_2)_3$/CN exhibited the highest CO-and CH_4- yields of 0.590 and 0.694 μmol/h, respectively.

The composites of S-doped CN (S-CN) with various amounts of Cu phytate (CuPA) (X=5, 10, 15, or 20 wt% of S-CN) were synthesized with 800 W microwave irradiation at 50°C, which were coded as $CuPA_X$/S-CN (X of 5–20 wt%), where Cu^{2+} was coordinated by phytic acid ligands [27]. The photocatalytic reduction of CO_2 was carried out by dispersing 10 mg catalyst powder in 20 mL distilled water under the irradiation with a $400 < \lambda < 780$ nm light from a 300 W Xe arc lamp. The results are shown in Figure 4.11a, revealing that the products from CO_2 reduction are principally CH_3OH with a small amount of CO. With the increase of CuPA-loading on the catalyst, the CH_3OH-yield increases initially up to 15% CuPA and then decreases. The $CuPA_{15}$/S-CN composite exhibited the CH_3OH-yield of 11.87 μmol/g·h, which was 9 times greater than that of the pristine S-CN. Experimentally, with the absence of light and the catalyst, no CO or CH_3OH could be detected. As shown in Figure 4.11b, the catalyst $CuPA_{15}$/S-CN shows a slight decrease in CH_3OH-yield after four cycles.

The *mpg*-CNs with different pore structure parameters were prepared from a mixture of cyanamide with different silica-colloid templates (0.2–1.5 wt%), onto which a ruthenium complex, trans(Cl)-[Ru{4,4′-$(CH_2PO_3H_2)_2$-2,2′-bipyridine}$(CO)_2Cl_2$], was loaded by dispersing in a methanol solution of the Ru-complex at room temperature to obtain the composite of *mpg*-C_3N_4 with the Ru complex. The resultant composites were able to convert CO_2 to formic acid (HCOOH) selectively [28–30]. The HCOOH-yields from the composite catalysts with different pore structure parameters are listed in Table 4.2.

NiO nanoparticles-loaded *g*-CN after wrapping with different amounts (1, 3, 5, 7, and 10 wt%) of single-layer rGnO, coded as $(rGnO)_X$/NiO/*g*-CN, where X corresponds

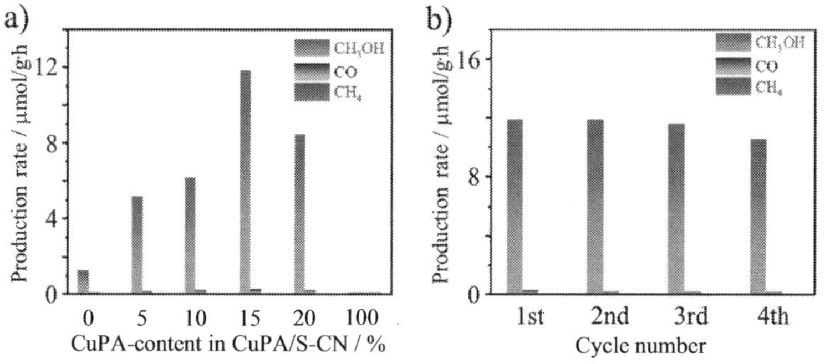

FIGURE 4.11 The CO_2-reduction performances of the CuPA-loaded S-CN composites ($CuPA_X$/S-CN): (a) in comparison with the pristine S-CN and CuPA under 300 W light irradiation and (b) cycle stability of $CuPA_{15}$/S-CN catalyst [27].

TABLE 4.2
Formic Acid Produced through CO_2 Reduction under Visible Light ($\lambda > 400$ nm) Using Different *mpg*-CN-Based Catalysts Loaded Ru-Complex by 3.9 μmol/g [29]

	Preparation Conditions		Pore Structure Parameters			
Code	SiO_2 Size (nm)	SiO_2/ Cyanamide	S_{BET} (m^2/g)	Pore Volume (cm^3/g)	Pore Diameter (nm)	Formic Acid Yield (nmol)
g-CN	–	–	6.1	–	–	Trace
CN-12-0.2	12	0.2	54	0.20	12.2	1,127
CN-12-0.5	12	0.5	105	0.37	12.2	1,854
CN-12-1.0	12	1.0	200	0.64	10.7	1,561
CN-12-1.5	12	1.5	240	0.74	10.7	Trace
CN-7-1.0	7	1.0	241	0.51	7.2	Trace
CN-24-1.0	24	1.0	130	0.70	24.4	1,223

to 1–10 wt% rGnO, were synthesized and exhibited stable CO_2-conversion performances [31]. 20 mg of the composite powder was dispersed in a mixture of 20 mL of deionized water containing 10 mL of TEOA after saturation with CO_2 under the irradiation of a 300 W Xe lamp. The production rates of CO and CH_4 and their cyclability on the $rGnO_5$/NiO/CN composite are shown in Figure 4.12a and b, respectively. The CO_2-conversion performances depend strongly on rGnO content, 5 wt% being the optimum ($rGnO_5$/NiO/CN), the yields of CO and CH_4 reaching 68.76 and 15.75 μmol/g, respectively, after 3 h of irradiation and having a high cyclability for CO_2 conversion.

Urchin-like α-Fe_2O_3 was loaded onto *g*-CN via hydrothermal treatment of aqueous suspension of Fe_2O_3 and *g*-CN at 150°C for 4 h to form α-Fe_2O_3/CN composite [32] and CO_2-conversion performances were studied. CO generation rate of 27.2 μmol/g·h was obtained without any co-catalyst and sacrifice reagent, which is >2.2 times higher than that of the pristine CN (10.3 μmol/g·h).

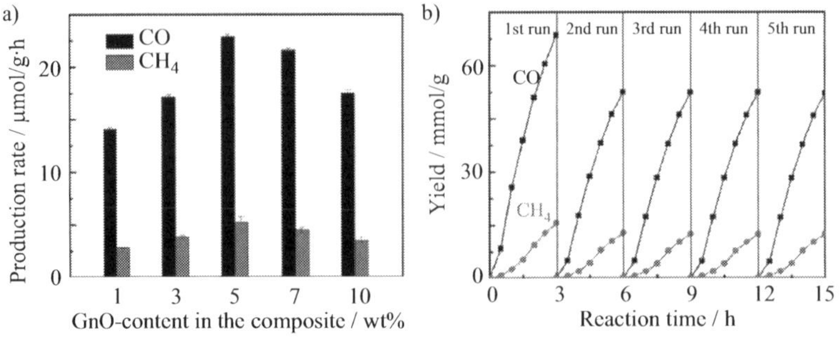

FIGURE 4.12 Photocatalytic CO_2-conversion performances of 5 wt% rGnO-coated NiO/CN composite photocatalysts $(rGnO)_5/NiO/CN)$: (a) production rates of CO and CH_4 and (b) cyclabilities [31].

The composites of *g*-CN with BiOBr being rich in surface oxygen vacancies (BiOBr-V_O) were synthesized and their CO_2 reduction and NO oxidation removal performances were studied [33]. The resultant composite (BiOBr-V_O)/CN exhibited the evolution rates of CO and CH_4 of 61.8 and 27.1 μmol/h·g, respectively, higher than the pristine CN (29.5 and 13.1 μmol/h·g, respectively) and BiOBr-V_O. The composite (BiOBr-V_O)/CN delivered a high selectivity for both CO and CH_4 of 96%. The important contribution of surface oxygen vacancies V_O to photocatalytic activities was shown in $Bi_{24}O_{31}Cl_{10}$ [34].

The composites of *g*-CN with $Bi_2O_2(NO_3)(OH)$, coded as BON/CNs, were synthesized and their CO_2-conversion performances were studied [35]. The composites were prepared by mixing the aqueous suspensions of melamine-derived *g*-CN and BON in the mass ratios of BON/*g*-CN of 1/9, 2/8, 3/7, and 4/6, kept stirring for 5 h and followed by centrifugation, washing, and drying at 60°C. The resultant composites are coded as BON_X/CN, where *X* corresponds to the BON/*g*-CN mass ratio (1, 2, 3, and 4). The photocatalytic CO_2 reduction experiment was conducted in a sealed quartz reactor using 20 mg catalyst powder and 3 mL deionized water by injecting CO_2 gas under the light from a 300 W Xe-lamp. Compositing of *g*-CN with BON accelerates the photocatalytic CO_2 reduction markedly, CO-generation rate being enhanced from 0.94 μmol/g·h h for the pristine BON and 3.29 μmol/g·h for the pristine *g*-CN up to 6.55, 14.84, 5.18, and 4.38 μmol/g·h for the composite BON_1/CN, BON_2/CN, BON_3/CN, and BON_4/CN, respectively.

A composite of *g*-CN with ultra-thin carbon quantum dot (CQD) nanosheets was synthesized by hydrothermal treatment of the mixture of melamine (5g) with CQDs (20 mg) in 50 mL of water at 180°C for 12 h and annealing at 660°C for 6 h in N_2 [36]. CQDs were prepared from an acetonitrile solution containing 50 mg o-phenylenediamine (OPD) and 25 mg 2,3-diaminomaleonitrile (DAMN) by heating at 180°C for 12 h, filtering (0.22 nm), and freeze-drying. CQD/CN composite aqueous suspension was frosted onto a 7 cm diameter round glass sheet, which was then contacted with high-purity CO_2 gas purged for 20 min (flow rate of 115 mL/min) under the irradiation of light ($\lambda > 420$ nm) from a 300 W Xe-lamp for 10 h. CO_2 reduction product is mainly CO with a minor amount of CH_4, CO yield reaching 120 μmol/g after 10 h irradiation with CO selectivity of 92.8%, while 39.2 and 2.0 μmol/g for the pristine CN and

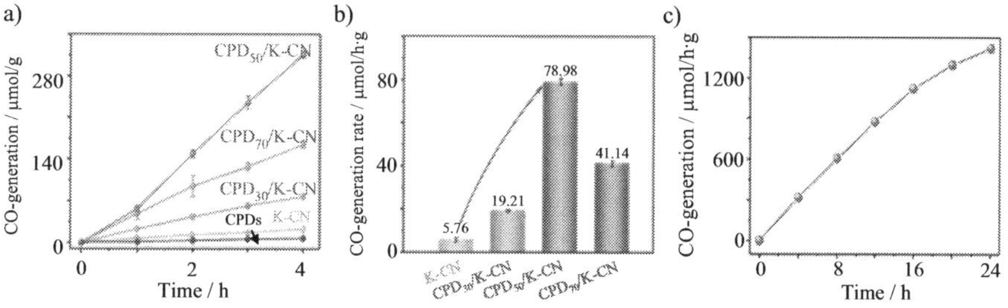

FIGURE 4.13 CO_2-reduction performances of CPD_X/K-CN composites: (a) CO production with time, (b) CO-production rate, and (c) CO-production stability for CPD_{50}/K-CN composite [37].

CQDs, respectively. The isotopically labeled $^{13}CO_2$ experiments confirmed that the photocatalytic product CO originated from $^{13}CO_2$ rather than the catalyst itself.

The composites of K-doped CN with carbonized polymer dots (CPDs), coded as (CPD_X/K-CN), were synthesized and studied on CO_2-reduction performances [37]. The composites were prepared by mixing K-CN (1 g) with different amounts (*X*) of CPDs (30, 50, and 70 mg) in 1 mL of ethanol, ball-milling, and then drying to obtain the composites CPD_X/K-CNs. K-CN was prepared by mixing melamine-derived *g*-CN (1 g) with 5.5 g KCl and 4.5 g LiCl, followed by calcination at 550°C, and PCDs were obtained by heating a deionized water solution of 5 mmol citric acid and 335 μL ethylenediamine at 200°C for 5 h. CO_2-reduction performances of the composites CPD_X/K-CNs are shown, focusing on the CO-generation in Figure 4.13a–c. On the resultant composites, the CO_2-reduction product was predominantly CO and its selectivity is estimated to be 100%. The CPD_{50}/K-CN composite exhibited the highest performance, with a rate of 78.98 μmol/h·g, being about 14 times larger than the pristine K-CN (5.76 μmo/h·g). This composite exhibited stable CO-generation for up to 24 h.

Composites of an O-doped *g*-CN with cobalt porphyrin (CoTPP) were synthesized ($CoTPP_X$/O-CN) [38]. O-CN was prepared from a mixture of urea (10 g) with ammonium formate (0.3 g) by calcination in an alumina crucible with lid at 550°C for 2 h. The resulting O-CN was dispersed in ethanol, followed by the addition of CoTPP at various CoTPP/O-CN mass ratios and stirred for 24 h under seal. The obtained solids were then heated at 250°C for 1 h in an alumina crucible to yield $CoTPP_X$/O-CN composites, where *X* represents the mass ratio of CoTPP to O-CN (0.5, 0.7, and 0.9 wt%). CO_2-reduction performances were measured by dispersing 10 mg of the catalysts in 12 mL water containing TEOA at 5°C and 80 kPa pressure of CO_2 under the light from a 300 W Xe-lamp. In Figure 4.14a, photoluminescence spectra of the composites $CoTPP_X$/O-CNs are compared with the pristine CN, revealing that the fluorescence intensity of $CoTPP_{0.7}$/O-CN was lower than that of the pristine CN, suggesting a better charge separation efficiency. The generation rates of the products of CO_2-photoreduction reactions (CO, CH_4, and H_2) are compared on the resultant catalysts in Figure 4.14b, demonstrating that compositing of CN with CoTPP is effective to enhance CO_2 reduction. The principal product was CO, associated with a small amount of CH_4, while H_2 was not detectable. The generations of CO and

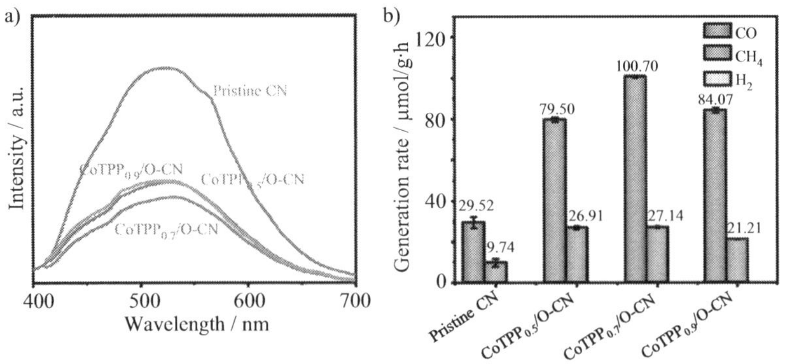

FIGURE 4.14 The CoTPP-loaded O-doped CN composites (CoTPP$_X$/O-CNs): (a) photoluminescence spectra and (b) the generation rates of the products from CO_2 photoreduction (CO, CH_4, and H_2) [38].

CH_4 gradually increased with the increase of CoTPP-loaded, and reached the highest at 0.7%, i.e., CoTPP$_{0.7}$/O-CN, giving 100.70 and 27.14 μmol/g·h, respectively.

A composite of *g*-CN with a zeolite-like porous boron imidazolate framework BIF-20 (coded as BIF-20/CN) exhibited an enhanced photoactivity for CO_2 reduction [39].

4.3 CONCLUDING REMARKS

Considerable enhancement in CO_2 adsorption capacity by moderate *g*-CN coating on highly microporous carbon foam (CF) was reported [40]. CF was prepared by a one-step pyrolysis of sodium ethoxide in a high flow rate of N_2 (6 L/min). The resultant CF was coated by *g*-CN through an impregnation into a solution of DCDA in a DCDA/CF weight ratio of 1/2 and stirring for either 1 or 24 h, followed by polymerization at 700°C for 2 h to form *g*-CN. The resultant CN-coated CFs are denoted as CN/CF-1 and -24 on the basis of the impregnation time of 1 and 24, giving different amounts of CN coating. Detailed pore structure parameters on the pristine CF with and without CN-coating in Table 4.3, revealing marked decreases in pore volumes. On these three adsorbents, CO_2 uptakes under 100 kPa and 2.0 MPa were measured at 273, 298, and 323 K, and the results are shown in Table 4.4. Under a low pressure of 100 kPa, CO_2 uptake increases after coating with *g*-CN, while a large amount of coating of *g*-CN made CO_2 uptake smaller. In this work [40], regrettably, photocatalytic activity seems not to be taken into consideration, probably because the authors are concerned only with CO_2 adsorption capturing from pre- or post-combustion gas mixtures. If photocatalytic CO_2 conversion by *g*-CN-based materials is taken into consideration, however, this result suggests various possibilities for photocatalytic CO_2 conversion and its environmental remediation applications. The *g*-CN coating on a CF with macropores may make the coating process of *g*-CN much simpler, the control of coating *g*-CN amount much easier, improve solar light efficiency on the photocatalyst surface, and release of the resultant CO and/or hydrocarbon gases becomes easier.

TABLE 4.3
Pore Structure Parameters of CF and CN-Coated CFs Calculated from the Adsorption–Desorption Isotherms of N_2 at 77 K and CO_2 at 273 K [40]

		Pore Volume (cm^3/g)				Pore Size (nm)			
Adsorbent	S_{BET} (m^2/g)	V_{ultra}[a]	V_{micro}[a]	V_{meso}	V_{total}	Ultra-Micropores[a]	Super-Micropores[a]	Micro-pores	Meso-pores
Pristine CF	2,537	0.4617	1.3407	0.2664	1.6071	0.40	0.63	1.19	2.46
CN/ CF-1	2,127	0.4495	1.1011	0.1940	1.2951	0.41	0.60	1.21	2.46
CN/CF-24	1,263	0.2774	0.6869	0.1116	0.7985	0.41	0.64	1.33	2.46

[a] Calculated from CO_2 adsorption–desorption isotherm at 273 K.

TABLE 4.4
CO_2 Uptake of CF and CN-Coated CFs (CN/CF-1 and -24) [40]

	CO_2 Uptake at 273 K (mmol/g)		CO_2 Uptake at 298 K (mmol/g)		CO_2 Uptake at 323 K (mmol/g)	
Adsorbent	100 kPa	2.0 MPa	100 kPa	2.0 MPa	100 kPa	2.0 MPa
Pristine CF	5.09	25.52	2.90	18.44	1.70	14.09
CN/CF-1	5.24	21.84	3.16	15.70	1.97	12.04
CN/CF-24	2.83	11.92	1.57	8.19	0.99	6.40

The photocatalytic reduction of CO_2 was discussed by density functional theory calculations on single-atom Pd and Pt-doped *g*-CN, i.e., (Pd1)-CN and (Pt1)-CN, respectively [41]. During CO_2 reduction, the individual metal atoms Pd and Pt located in the intrinsic cavity of V_{6N} of *g*-CN function as the active sites, while the substrate *g*-CN provides the source of hydrogen (H*) through its H_2-evolution reaction. As a consequence, the (Pd1)-CN and (Pt1)-CN are expected to exhibit excellent activity in CO_2 reduction, yielding preferentially HCOOH on the (Pd1)-CN, while the (Pt1)-CN prefers to reduce CO_2 to CH_4. The position of (Cu1) in P-CN was reported to govern the product; (Cu1) at 3N-coordinated intrinsic cavity produces mainly CO, while (Cu1) at the 3P-coordinated position produces mainly H_2, with a small amount of CO [19].

In Table 4.5, CO_2-conversion (CO_2-reduction) performances are summarized by focusing mostly CO- and CH_4-generation rate with main characteristics of the *g*-CN-based catalysts. As pointed out on H_2-evolution (**Chapter 3**), the crystallinity of *g*-CN is supposed to be an important factor for CO_2-conversion (CO_2-reduction) performance, but studies on highly crystalline *g*-CN have been reported only rarely. The *g*-CN synthesized using KCl-LiCl molten salt was used for doping of single Pt (Pt1) and Cu (Cu1) atoms [23] and for the preparation of the composite with CPDs, which delivered a high CO-generation rate of about 79 μmol/g·h rate [37]. The pristine *g*-CN used in this paper is reported to have a CO-generation rate of 5.76 μmol/g·h. *g*-CN, which was used to load CQDs (CQD/CN composite), also delivered a

TABLE 4.5
CO_2-Conversion (CO_2-Reduction) Performances of *g*-CN-Based Photocatalysts

			CO_2 Reduction				
			Conditions	CO_2-Reduction Product and Rate			
Photo-Catalyst	Synthesis Conditions	*g*-CN Character	Co-Catalyst, Scavenger, etc.	CO (μmol/h·g)	CH_4 (μmol/h·g)	Others	References
			***g*-C_3N_4 and Modified *g*-C_3N_4**				
V_{N3C}-CN-50	Urea, 50 vol% glycerol + water	Containing N-vacancy (V_{N3C})	Catalysis on a quartz sand plate, contact with CO_2 gas at 6°C,	4.18	–	–	11
V_N-CN-5	Light irradiation for 5 h	Control of N-vacancy (V_N)	Loading a catalyst on a glass-fiber paper	35.2	–	–	12
$B_{3.0}$-CN	3.0 at% B-doping (BH_3NH_3)		Au, Electrochemical	–	–	C_2H_5OH	14
$P_{0.4}C$-CN	P- and C-co-doping		TEOA	Trace	41.84	H_2 Trace	15
PB-*ut*-CN	P- and B-co-doping	Ultrathin CN (*ut*-CN)	Catalysis on a quartz glass plate, contact with CO_2 gas at r.t.	22.45	–	–	16
$(Cu1)_{0.5}$-V_{N2C}-CN	Single Cu atom (Cu1)-doping	Cu1-doping, V_{N2C}-introduction	No co-catalyst No scavenger	11.12	Trace	–	17
(Co1)(Mn1)-CN	(Co1)-(Mn1)-co-doping			47	–	–	18

(*Continued*)

TABLE 4.5 (*Continued*)
CO_2-Conversion (CO_2-Reduction) Performances of *g*-CN-Based Photocatalysts

			CO_2 Reduction				
			Conditions	CO_2-Reduction Product and Rate			
Photo-Catalyst	Synthesis Conditions	*g*-CN Character	Co-Catalyst, Scavenger, etc.	CO (µmol/h·g)	CH_4 (µmol/h·g)	Others	References
$(Cu1)_{3N}$/P-CN	(Cu1) at 3N-coordination	P-doped CN (P-CN)	In a CO_2 atmosphere (1 atm, 20°C)	49.8	–	–	19
$(Cu1)_{3P}$/P-CN	(Cu1) at 3P-coordination			3.7	–	H_2 62.5 µmol/g·h	
(Ni1)-S-CN-N	(Ni1) and S-co-doping	NH_4Cl-treatment	In 80 kPa CO_2 gas	58.6	18.9	–	20
$(Er1)_{6N}$/*nt*-CN	20.1 wt% (Er1)-doping	*g*-CN nanotube (*nt*-CN)	In pure-water	47.1	2.5	–	21
(Ag1)/CN	(Co1) and (Ag1) co-doping	P-doped CN (P-CN)	In CH_3CN/H_2O (6/4 by volume)	4.38[a]	1.57[a]	–	22
(Co1) (Ag1)/P-CN				5.38[a]	–	–	
(Co1) (Ag_{1+n})/P-CN				11.71[a]	–	–	
(Pt1)(Cu1)/*cr*-CN	(Pt1)-(Cu1)-co-doping	Highly crystalline *g*-CN (*cr*-CN)	–	11.7[a]	2.8[a]	–	23
(Ni1)/CN-Ar-2	(Ni1)-doping at 550°C in Ar atmosphere for 2 h	Ni-vapor exfoliated CN	–	19.85	1.73	–	24

(*Continued*)

TABLE 4.5 (*Continued*)
CO_2-Conversion (CO_2-Reduction) Performances of *g*-CN-Based Photocatalysts

			CO_2 Reduction				
			Conditions	CO_2-Reduction Product and Rate			
Photo-Catalyst	Synthesis Conditions	*g*-CN Character	Co-Catalyst, Scavenger, etc.	CO (μmol/h·g)	CH_4 (μmol/h·g)	Others	References
(Ni1)-OB-*ut*-CN	(Ni1) and borate-doping	Ultrathin *g*-CN (*ut*-CN)	No sacrificial reagents	22.1	8.7	–	25
		***g*-C_3N_4-Based Composites**					
$(CeO_2)_2$/CN	2 wt% CeO_2-loading	*mpg*-CN SiO$_2$-templating	Dispersed on a flat glass plate contacting with mixture of CO_2 and water vapor	11.8[a]	9.08[a]	–	26
$(CeO_2)_3$/CN	3 wt% CeO_2-loading			10.2[a]	13.88[a]	–	
CuPA$_x$/S-CN	Cu phytate (CuPA)-doping	S-doped CN (S-CN)	10 mg catalyst with 20 mL of water	Small	Trace	CH_3OH 11.87 μmol/g·h	27
Ru-complex-loaded *g*-CN	*Trans*(Cl)-[Ru{4,4′-$(CH_2PO_3H_2)_2$-2,2′-bipyridine}$(CO)_2Cl_2$]	*mpg*-CN	Dispersing in an acetonitrile with 20 vol% TEOA	–	–	HCOOH 1.854 mmol	29
5 wt% rGnO-wrapped NiO/*g*-CN	Wrapping by single-layered GnO	NiO-loaded *g*-CN	20 mg catalyst with 20 mL water containing TEOA	22.92[a]	5.25[a]	–	31
α-Fe_2O_3/CN	Urchin-like α-Fe_2O_3		No co-catalyst No scavenger	27.2	–	–	32

(*Continued*)

TABLE 4.5 (*Continued*)
CO_2-Conversion (CO_2-Reduction) Performances of *g*-CN-Based Photocatalysts

			CO_2 Reduction				
			Conditions	CO_2-Reduction Product and Rate			
Photo-Catalyst	Synthesis Conditions	*g*-CN Character	Co-Catalyst, Scavenger, etc.	CO (μmol/h·g)	CH_4 (μmol/h·g)	Others	References
BiOBr-V_O/CN	BiOBr with surface O-vacancies (V_O)			61.8	27.1	–	33
BON/CN 2/8 by mass	$Bi_2O_2(NO_3)(OH)$ (BON)-loading		20 mg catalyst in a Petri dish 15 mL water	14.84	–	–	35
CQD/CN	Carbon quantum dot (CQD)-loading	Melamine at 660°C for 6h	5 mg catalyst on a glass plate + 30 mL water	48	–	–	36
CPD_{50}/K-CN	Carbonized polymer dot (CPD)-loading	K-doped CN synthesized by KCl-LiCl	At 5°C, 30 mg catalyst + 50 mL water, CO_2 bubbling	78.98	–	–	37
$CoTPP_{0.7}$/O-CN	Cobalt porphyrin (CoTPP)-loading	O-doped CN (O-CN)	10 mg catalyst + 12 mL water with TEOA, 80 kPa CO_2	100.70	27.14	–	38

[a] Recalculated to the common unit (μmol/g·h) by taking into account the amount of catalyst used.

relatively high CO-generation rate 48 μmol/g·h [36]. The support (*g*-CN), which was synthesized from melamine at relatively high temperature (660°C for 6 h) delivered a rate of 15.7 μmol/g·h. CQDs used in these two papers might be a key material for CO_2-reduction reaction under visible light, while CPDs and CQDs themselves did not deliver a high CO-generation rate (1.85 and 2.0 μmol/g·h). On carbon dots (CPDs and CQDs), the reproducibility and stability of CO_2-reduction performance must be studied in detail to improve them.

Several literatures reported that doping of single metal atoms to *g*-CN achieved an effective enhancement in CO_2-conversion performance, as shown in Table 4.5. Considering the fact that "metal-free" is one of the characteristics (or advantages) of the photocatalyst *g*-CN, the introduction of metal atoms into *g*-CN must be discussed deeply. However, it has to be pointed out that the *g*-CNs doped by single metal atoms, such as (Cu1) and (Ni1), could produce low-molecular-weight hydrocarbons (CH_4 and CH_3OH) in addition to CO, whereas the most reported *g*-CN-based catalysts for CO_2 reduction generated CO.

REFERENCES

1. Yui T, Kan A, Saitoh C, et al., Photochemical reduction of CO_2 using TiO_2: Effects of organic adsorbates on TiO_2 and deposition of Pd onto TiO_2. *ACS Appl Mater Interfaces* 2011; 3: 2594–600.
2. Iizuka K, Wato T, Miseki Y, et al., Photocatalytic reduction of carbon dioxide over Ag cocatalyst-loaded $ALa_4Ti_4O_{15}$ (A=Ca, Sr, and Ba) using water as a reducing reagent. *J Am Chem Soc* 2011; 133: 20863–8.
3. Teramura K, Iguchi S, Mizuno Y, et al., Photocatalytic conversion of CO_2 in water over layered double hydroxides. *Angew Chem Int Ed* 2012; 51: 8008–11.
4. Yin H, Dong F, Wang D, et al., Coupling Cu single atoms and phase junction for photocatalytic CO_2 reduction with 100% CO selectivity. *ACS Catal* 2022; 12(22): 14096–105.
5. Li C, Lu X, Chen L, et al., WO_3/BiOBr S-scheme heterojunction photocatalyst for enhanced photocatalytic CO_2 reduction. *Materials* 2024; 17(13): 3199.
6. Liu X, Chen T, Xue Y, et al., Nanoarchitectonics of MXene/semiconductor heterojunctions toward artificial photosynthesis via photocatalytic CO_2 reduction. *Coord Chem Rev* 2022; 459: 214440.
7. Kumar A, Hasija V, Sudhaik A, et al., Artificial leaf for light-driven CO_2 reduction: Basic concepts, advanced structures and selective solar-to-chemical products. *Chem Eng J* 2022; 430: 133031.
8. Samanta S, Srivastava R, Catalytic conversion of CO_2 to chemicals and fuels: The collective thermocatalytic/photocatalytic/electrocatalytic approach with graphitic carbon nitride. *Mater Adv* 2020; 1: 1506–45.
9. Alekseev RF, Saraev AA, Kurenkova AY, et al., Heterostructures based on *g*-C_3N_4 for the photocatalytic CO_2 reduction. *Russ Chem Rev* 2024; 93(5): 5124.
10. Xia P, Antonietti M, Zhu B, et al., Designing defective crystalline carbon nitride to enable selective CO_2 photoreduction in the gas phase. *Adv Funct Mater* 2019; 29(15): 1900093.
11. Gong Y, Yang P, Ma D, et al., Construction of three-coordinated (N_{3C}) nitrogen vacancies in *g*-C_3N_4 for efficient photocatalytic CO_2 reduction. *Ceram Int* 2024; 50(18B): 33131–42.
12. Li L, Liu H, Cheng C, et al., Photochemical tuning of tricoordinated nitrogen deficiency in carbon nitride to create delocalized π electron clouds for efficient CO_2 photoreduction. *ACS Catal* 2024; 14: 10204–13.

13. Gong Y, Chen J, Ma D, Zhong J, Green preparation of nitrogen vacancies enriched *g*-C_3N_4 for efficient photocatalytic reduction of CO_2 and Cr (VI). *J Colloid Interface Sci* 2025; 682: 446–59.
14. Sagara N, Kamimura S, Tsubota T, et al., Photoelectrochemical CO_2 reduction by a p-type boron-doped *g*-C_3N_4 electrode under visible light. *Appl Catal B* 2016; 192: 193–8.
15. Huang QS, Li Q, Chu C, et al., Synergetic regulation of electronic structure of graphitic carbon nitride through phosphorus and carbon co-doping for enhanced photocatalytic CO_2 reduction. *Chem Eng J* 2024; 482: 149155.
16. Hussien MK, Sabbah A, Qorbani M, et al., Constructing B–N–P bonds in ultrathin holey *g*-C_3N_4 for regulating the local chemical environment in photocatalytic CO_2 reduction to CO. *Small* 2024; 20(35): 2400724.
17. Duan Y, Wang Y, Zhang W, et al., Simultaneous CO_2 and H_2O activation via integrated Cu single atom and N vacancy dual-site for enhanced CO photo-production. *Adv Funct Mater* 2023; 33: 2301729.
18. Ou H, Ning S, Zhu P, et al., Carbon nitride photocatalysts with integrated oxidation and reduction atomic active centers for improved CO_2 conversion. *Angew Chem Int Ed* 2022; 61(34): e202206579.
19. Sun X, Sun L, Li G, et al., Phosphorus tailors the *d*-band center of copper atomic sites for efficient CO_2 photoreduction under visible-light irradiation. *Angew Chem Int Ed* 2022; 61(38): e202207677.
20. Xu J, Chen Y, Chen M, et al., *In situ* growth strategy synthesis of single-atom nickel/sulfur co-doped *g*-C_3N_4 for efficient photocatalytic tetracycline degradation and CO_2 reduction. *Chem Eng J* 2022; 442: 136208.
21. Ji S, Qu Y, Wang T, et al, Rare-earth single erbium atoms for enhanced photocatalytic CO_2 reduction. *Angew Chem Int Ed* 2020; 59: 10651–7.
22. Deng A, Zhao E, Li Q, et al, Atomic cobalt-silver dual-metal sites confined on carbon nitride with synergistic Ag nanoparticles for enhanced CO_2 photoreduction. *ACS Nano* 2023; 17: 11869–81.
23. Cheng L, Zhang P, Wen Q, et al., Copper and platinum dual-single-atoms supported on crystalline graphitic carbon nitride for enhanced photocatalytic CO_2 reduction. *Chin J Catal* 2022; 43: 451–60.
24. Ma L, Guan R, Kang W, et al., Preparation of highly dispersed Ni single-atom doped ultrathin *g*-C_3N_4 nanosheets by metal vapor exfoliation for efficient photocatalytic CO_2 reduction. *J Colloid Interface Sci* 2024; 660: 381–92.
25. Wang Y, Qu Y, Qu B, et al., Construction of six-oxygen-coordinated single Ni sites on *g*-C_3N_4 with boron-oxo species for photocatalytic water-activation-induced CO_2 reduction. *Adv Mater* 2021; 33: 2105482.
26. Li M, Zhang L, Wu M, et al., Mesostructured CeO_2/*g*-C_3N_4 nanocomposites: Remarkably enhanced photocatalytic activity for CO_2 reduction by mutual component activations. *Nano Energy* 2016; 19: 145–55.
27. Du W, Chen R, Yang L, et al., Phytic acid-bridged copper on sulfur-containing carbon nitride for enhancing photocatalytic CO_2 reduction to CH_3OH. *Ind Eng Chem Res* 2024; 63(14): 6201–9.
28. Maeda K, Sekizawa K, Ishitani O, A polymeric-semiconductor–metal-complex hybrid photocatalyst for visible-light CO_2 reduction. *Chem Commun* 2013; 49: 10127–9.
29. Maeda K, Kuriki R, Zhang X, et al., The effect of the pore-wall structure of carbon nitride on photocatalytic CO_2 reduction under visible light. *J Mater Chem A* 2014; 2: 15146–51.
30. Kuriki R, Sekizawa K, Ishitani O, et al., Visible-light-driven CO_2 reduction with carbon nitride: Enhancing the activity of ruthenium catalysts. *Angew Chem Int Ed* 2015; 54(8): 2406–9.

31. Tao FF, Dong Y, Yang L, NiO/*g*-C_3N_4 p–n heterojunctions wrapped by rGO for the enhanced CO_2 photocatalytic reduction. *ACS Sustain Chem Eng* 2024; 12(17): 6709–18.
32. Jiang Z, Wan W, Li H, et al., A hierarchical Z-Scheme α-Fe_2O_3/*g*-C_3N_4 hybrid for enhanced photocatalytic CO_2 reduction. *Adv Mater* 2018; 30: 1706108.
33. Liu D, Chen D, Li N, et al., Surface engineering of *g*-C_3N_4 by stacked BiOBr sheets rich in oxygen vacancies for boosting photocatalytic performance. *Angew Chem Int Ed* 2020; 59: 4519–24.
34. Jin X, Lv C, Zhou X, et al., Oxygen vacancy engineering of $Bi_{24}O_{31}Cl_{10}$ for boosted photocatalytic CO_2 conversion. *ChemSusChem* 2019; 12: 2740–7.
35. Liu T, Hao L, Bai L, et al., Z-scheme junction $Bi_2O_2(NO_3)(OH)$/*g*-C_3N_4 for promoting CO_2 photoreduction. *Chem Eng J* 2022; 429: 132268.
36. Xu Y, Hou W, Huang K, et al., Engineering built-in electric field microenvironment of CQDs/*g*-C_3N_4 heterojunction for efficient photocatalytic CO_2 reduction. *Adv Sci* 2024; 34: 2403607.
37. Dong J, Zhao J, Yan X, et al., Construction of carbonized polymer dots/potassium doped carbon nitride nanosheets Van der Waals heterojunction by ball milling method for facilitating photocatalytic CO_2 reduction performance in pure water. *Appl Catal B-Environ Energy* 2024; 351: 123993.
38. Zheng Z, Zheng Q, Qiao B, et al., S-scheme heterojunction of thin-layer O-modified graphitic carbon nitride/cobalt porphyrin to promote photocatalytic CO_2 conversion. *J Colloid Interface Sci* 2024; 667; 713–72.
39. Xu G, Zhang H, Wei J, et al., Integrating the *g*-C_3N_4 nanosheet with B–H bonding decorated metal-organic framework for CO_2 activation and photoreduction. *ACS Nano* 2018; 12: 5333–40.
40. Vorokhta M, Kusdhany MIM, Švábová M, et al., Hierarchically porous carbon foams coated with carbon nitride: Insights into adsorbents for pre-combustion and post-combustion CO_2 separation. *Sep Purif Technol* 2025; 354(5): 129054.
41. Gao G, Jiao Y, Waclawik ER, et al, Single atom (Pd/Pt) supported on graphitic carbon nitride as an efficient photocatalyst for visible-light reduction of carbon dioxide. *J Am Chem Soc* 2016; 138: 6292–7.

5 Hydrogen Peroxide Production

Hydrogen peroxide (H_2O_2) has drawn increasing attention because of its strong oxidation power with an environmentally friendly nature by emitting only water as a byproduct. It has been widely used in the chemical industry, in sanitation, and in environmental remediation. It is also frequently employed in biomedicine, including sensing and cancer therapy, owing to its endogenous nature. However, it has mainly been produced by the anthraquinone process, which is environmentally unfriendly due to its high energy consumption [1]. Therefore, there is a strong desire for the development of an efficient, low-cost, and environmentally friendly process for H_2O_2 synthesis. Photocatalytic H_2O_2 production on semiconductors, such as TiO_2, has been studied extensively. g-C_3N_4 is successfully activated by the visible light irradiation to an efficient production of H_2O_2 according to the following equations.

$$O_2 + e^- \rightarrow \bullet O_2^- \tag{5.1}$$

$$O_2^- + e^- + 2H^+ \rightarrow H_2O_2 \tag{5.2}$$

5.1 g-C_3N_4 AND MODIFIED g-C_3N_4

g-CN was synthesized by the polymerization of different precursors, melamine (M), ammonium thiocyanate (ATC), urea (U), thiourea (TU), and dicyandiamide (DCDA) at 600°C [2], which were coded as X-CN, where X corresponds to the type of precursor. In addition, the mixtures of these precursors with KCl (melting point of 747°C) were also polymerized at 600°C to produce K-doped CN (coded as K-X-CN). The performance of H_2O_2 photoproduction was evaluated in a 5 vol.% ethanol solution (a scavenger), and the results are shown in Figure 5.1. The H_2O_2-photoproduction rate for K-M-CN, K-ATC-CN, K-U-CN, K-TU-CN, and K-DCDA-CN was measured as 18.5, 26.7, 14.1, 23.3, and 12.2 mmol/h·g, respectively. The doping of K^+ is effective for the enhancement of H_2O_2 production, for instance, the photocatalytically generated H_2O_2 in 1 h on K-ATC-CN and K-TU-CN was 3.70 and 2.44 mM, which was 23.2 and 19.1 times that on ATC-CN and TU-CN (without K-doping), respectively.

The photocatalytic production of H_2O_2 was performed using the cyanamide-derived g-C_3N_4 catalyst in an ethanol/water mixture (9/1 by volume) under a visible light irradiation from a Xe lamp ($\lambda > 420$ nm) with magnetic stirring under an O_2 atmosphere (1 atm) at 25°C [3]. The yields of the products, including H_2O_2, CH_3CHO, CH_3COOH, and CO_2, and the H_2O_2 selectivity are shown as the function of the reaction temperature and the water content in the solution in Figure 5.2a and b, respectively. Production yield of H_2O_2 reached 30 μmol from a system of 5 mL ethanol/water mixture with 20 mg catalyst, which was much higher than that using

DOI: 10.1201/9781003589051-5

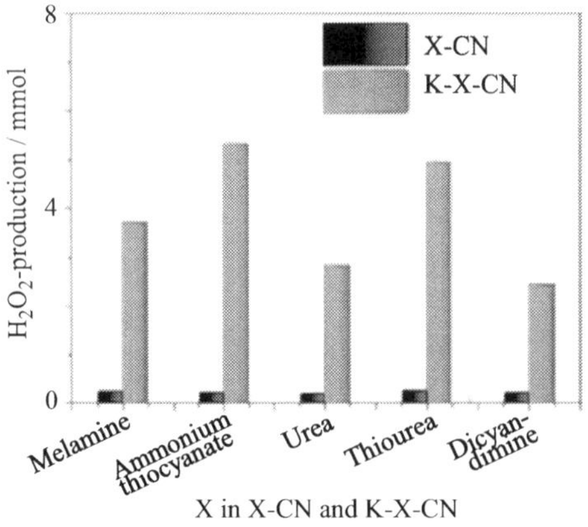

FIGURE 5.1 Photocatalysis performance of K-X-CN and X-CN for H_2O_2 production [2].

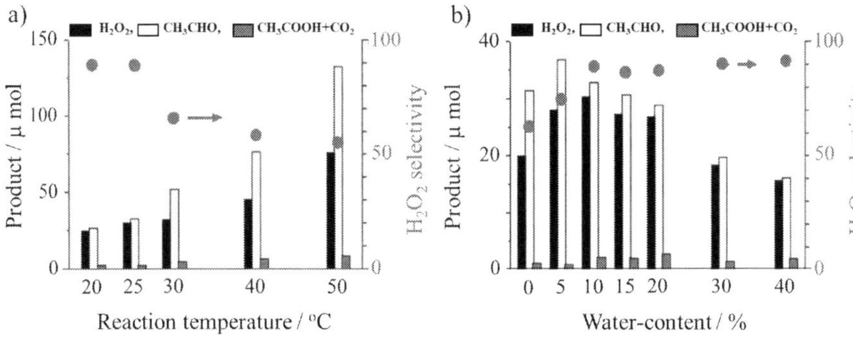

FIGURE 5.2 Yields of oxidation products (H_2O_2, CH_3CHO, and CH_3COOH) and H_2O_2 selectivity during photoreaction with g-C_3N_4 in an EtOH/water/O_2 system: effects (a) of reaction temperature and (b) of water-content in the solution [3].

other catalysts, such as TiO_2, Pt/TiO_2, Au/TiO_2, (<0.2 µmol yield), and even with $Pt_{0.1}$/g-C_3N_4 (4.2 µmol yield).

g-CN, which had an enhanced electron transfer and abundant active surfaces, was synthesized through suitable basicity-assisted thermal polymerization of urea (5 g) with the addition of KCl (5 g) and KOH (0.35 g) at 550°C for 4 h in a sealed crucible [4]. The resultant CNs exhibited a high H_2O_2-production rate of 200 µmol/g·h in pure water.

Amorphous CN (*a*-CN) was prepared by the treatment of the melamine/cyanuric acid supramolecular aggregates (MCSA) in H_2 plasma under a pressure of 150 Pa in a H_2 flow rate of 30 sccm and a temperature of 400°C for 90 min [5]. The resultant *a*-CN contained two N-vacancies at a 2C coordinated site V_{N2C} in one CN unit. H_2O_2 production was studied by dispersing 25 mg of photocatalyst in 50 mL of 10% isopropanol (IPA) solution with O_2 bubbling under the irradiation of the light from a 300 W Xe lamp. For comparison, *g*-CN was prepared from MCSA at 550°C and bulk *g*-CN

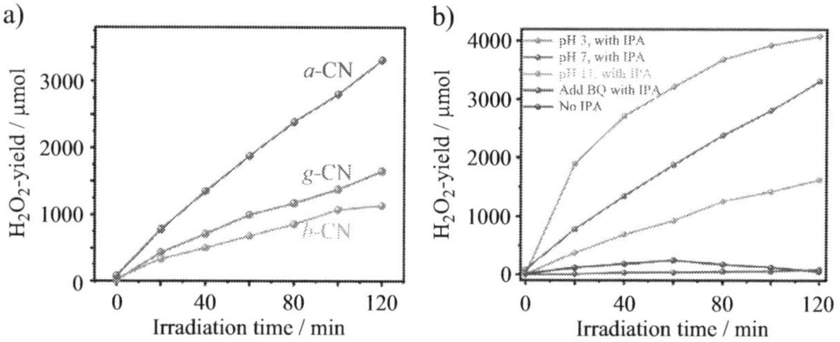

FIGURE 5.3 H_2O_2-production performances over *a*-CN: (a) H_2O_2-production kinetics in comparison with the pristine *g*-CN and bulk CN (*b*-CN) and (b) the influence of pH, IPA, and BQ during the H_2O_2 production on the *a*-CN under simulated solar irradiation (pH is adjusted by HCl and KOH) [5].

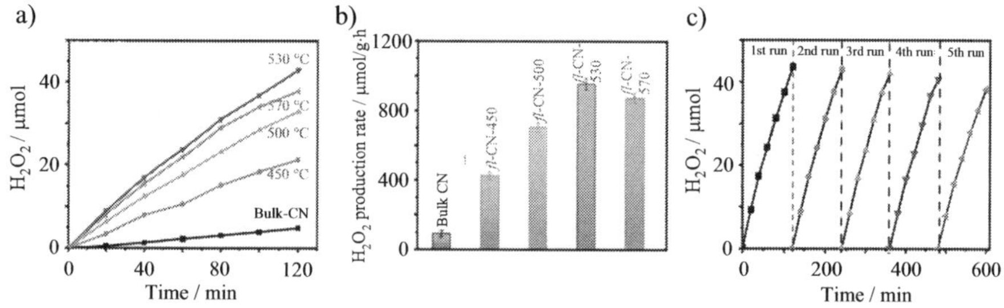

FIGURE 5.4 H_2O_2-production performances over the CN exfoliated at different temperatures (*fl*-CN-*X*): the effect of exfoliation temperature on (a) H_2O_2-production kinetics, (b) production rates, and (c) cyclability for H_2O_2 production on the *fl*-CN-530 [6].

(*b*-CN) from melamine at 550°C. In Figure 5.3a, H_2O_2-production kinetics over *a*-CN is shown in comparison with those of the pristine *g*-CN and *b*-CN, revealing high H_2O_2 production over *a*-CN with 3,312.2 μmol H_2O_2 yield after 2 h irradiation, whereas the pristine *g*-CN and *b*-CN gave 1,647.8 and 1,137.4 μmol, respectively. In Figure 5.3b, effects of solution pH and the addition of IPA and benzoquinone (BQ) scavenger reagent on H_2O_2-production kinetics are shown, revealing that the presence of IPA in the solution is enhanced markedly on the H_2O_2 production.

The few-layered *g*-CN nanosheets were prepared via liquid-exfoliation (nitrate-intercalation) of the *b*-CN derived from melamine at different temperatures of 450°C, 500°C, 530°C, and 570°C [6]. The resultant exfoliated few-layered CNs are coded as *fl*-CN-*X*, where *X* corresponds to the exfoliation temperature. The H_2O_2-production performances were studied by dispersing the catalyst (25 mg) in a solution containing 45 mL of distilled water and 5 mL of ethanol (scavenger), followed by O_2 bubbling at a rate of 50 mL/min under the irradiation from a 300 W Xe lamp with a UV-cutoff filter (420 nm $\leq \lambda \leq$ 780 nm). The results are shown in Figure 5.4a–c. The H_2O_2 production depends strongly on the exfoliation temperature of the *b*-CN, the *fl*-CN-530 giving the highest H_2O_2 production and rate reaching 952 μmol/g·h (Figure 5.4a and b). A calculation based on UV-vis diffuse reflectance spectra showed that the valence band potential E_{VB} was around 2.02 eV, while the conduction band potential E_{CB} decreased from −0.72 eV for the pristine *b*-CN to −0.86 and −0.92 V for the *fl*-CN-530 and 570, respectively. In other words, the bandgap increased with exfoliation from 2.74 eV for the pristine *b*-CN to 2.87 and 2.93 eV after the exfoliation at 530°C and 570°C, respectively. The resultant *fl*-CNs could photo-catalytically oxidize CH_4 by *on-site* formed H_2O_2.

CN was regenerated from a solution of *g*-CN and $ZnCl_2$ molten salt hydrates by dilution with ethanol (*r*-CN) [7]. This regeneration process resulted in an exfoliation of the pristine *g*-CN. Consequently, it became a nanobelt morphology (about 20 nm width and hundreds of nm length), C/N atomic ratio of 0.609, and bandgap of 2.69 eV. The concentration of H_2O_2 increases steadily with time, much more rapidly over *r*-CN than over the pristine *g*-CN, as shown in Figure 5.5a. After 1 h irradiation with 300 W Xe light, *r*-CN produced 18.5 mmol/L of H_2O_2, whereas the pristine *g*-CN only produced about 3 mmol/L under the same conditions. In Figure 5.5b, the H_2O_2-production rate under different light irradiation over *g*-CN and *r*-CN, revealing a marked acceleration of H_2O_2 production on *r*-CN.

Intercalation of black phosphorus (BP) into *g*-CN (BP-CN) was achieved in an NMP solution containing different amounts of BP [8]. During sonication, *g*-CN sheets were exfoliated (liquid exfoliation) to 1.0–1.7 nm thickness. The BP-CN containing 10% BP exhibited the highest H_2O_2-production rate of 540 μmol/g·h.

B-doped *g*-CN (B-CN) could accelerate the photocatalytic H_2O_2 production because the electron-deficient boron acts as Lewis acid sites, which could improve the light absorption ability and enhance H_2O_2 production through the effective activation of ethanol [9]. B-CN was synthesized from a mixture of polymeric CN with $NaBH_4$ by heat treatment at 250°C, 350°C, and 450°C for 1 h in N_2 flow (coded as B-CN-*X*, where *X* corresponds to the calcination temperature. In comparison, the B-CN-450 was mixed with either NaCl or additional $NH_3{\cdot}BH_3$ and then calcined to get Na-B-CN and B-B-CN. The H_2O_2 photocatalytic production behavior was investigated by dispersing the catalyst powder in 50 mL of either pure water or 10 vol%

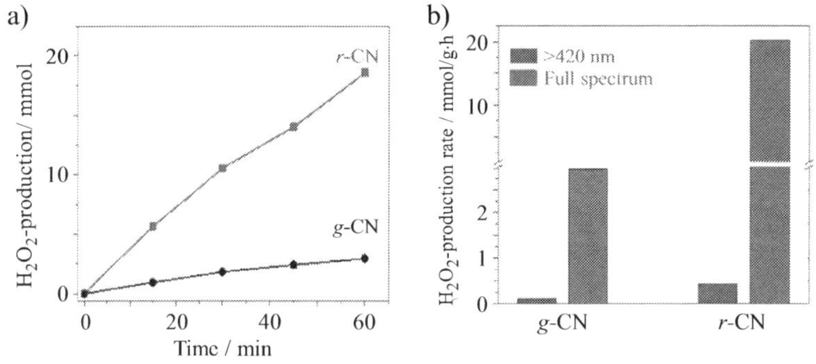

FIGURE 5.5 H_2O_2-production performances of regenerated CN (*r*-CN) from $ZnCl_2$ molten salt hydrate solution in comparison with the pristine *g*-CN: (a) H_2O_2-production kinetics and (b) H_2O_2-production rates under the light with >420 nm and in full spectrum [7].

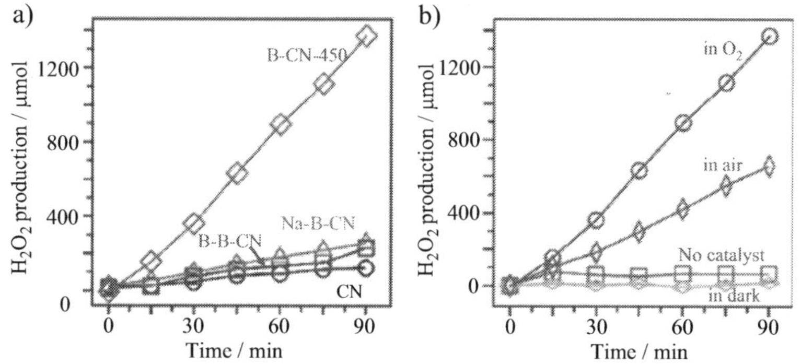

FIGURE 5.6 H_2O_2-production performances of B-CN-450: (a) time changes in H_2O_2 production in comparison with the pristine CN, Na-B-CN, and B-B-CN catalysts and (b) effect of atmospheres on B-CN-450 catalyst [9].

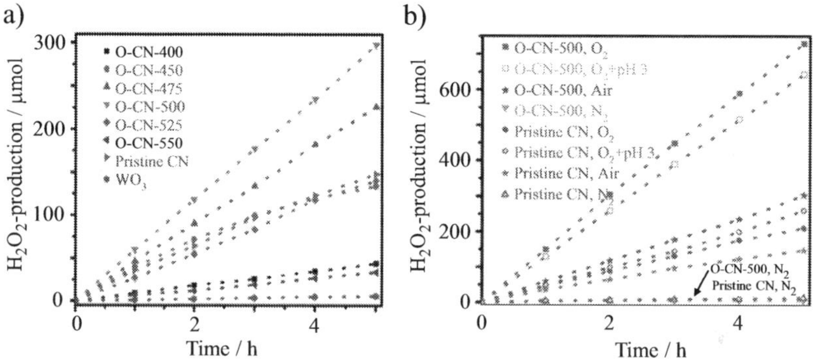

FIGURE 5.7 H_2O_2-production performances of O-enriched *g*-CN in comparison with WO_3: (a) H_2O_2-production kinetics in water with 10% isopropanol under visible light irradiation and an air atmosphere and (b) H_2O_2-production kinetics in water with 10% isopropanol under different atmospheres and pHs [10].

ethanol/water at 24°C under irradiation with a 300 W Xe lamp with a 420 nm-cutoff filter. In Figure 5.6a, the H_2O_2-production kinetics over the resultant B-CN prepared at 450°C (B-CN-450) are compared with Na-B-CN, B-B-CN, and the pristine CN. The B-CN-450 exhibited the highest photocatalytic activity for the H_2O_2 production, reaching 51,008 μmol/g·h, much higher than the pristine CN (4,113 μmol/g·h). The presence of O_2 in the atmospheric gas governs the H_2O_2 production, as shown in Figure 5.6b.

O-enriched *g*-CN was prepared by calcination of a mixture of 50 mmol of DCDA with 1 mol% ammonium para-tungstate in a lidded crucible at 400°C, 450°C, 475°C, 500°C, 525°C, and 550°C for 4 h [10]. The resultant CNs are coded as O-CN-*X*, where *X* corresponds to calcination temperature. H_2O_2-production kinetics of the resultant O-CN-*X* are shown in comparison with the pristine CN and WO_3 obtained by calcination of ammonium para-tungstate at 500°C in Figure 5.7a. The catalyst

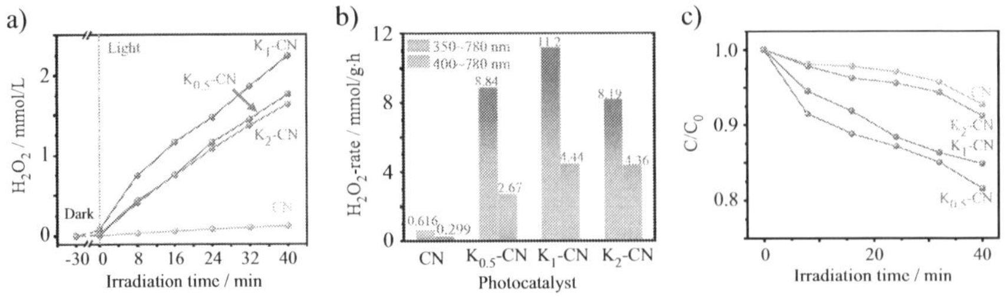

FIGURE 5.8 K^+-intercalated *g*-CNs (K_X-CNs) and the pristine CN: (a) H_2O_2-production curves, (b) H_2O_2-production rate under the light with different wavelengths, and (c) H_2O_2-decomposition curves with K-CNs catalysts [11].

O-CN-500 exhibited the highest H_2O_2-production performance, with the rate reaching 58.4 mmol/g·h in water containing 10% IPA. In Figure 5.7b, the H_2O_2-production kinetics under different atmospheres and pHs on O-CN-500, revealing the importance of a high concentration O_2 atmosphere.

It has been shown that K^+-intercalation into crystalline *g*-CN (K-CN) effectively enhances H_2O_2 production [11]. The K-CNs synthesized from the mixtures of urea (20 g) with varying amounts of KCl (0.5, 1, and 2 g) are coded as K_X-CN, where X corresponds to the amount of KCl used (X is 0.5, 1, and 2). The doped K^+ was supposed to be intercalated into the intrinsic cavity formed by C-N linkages between triazine moieties. The resultant catalyst powder was dispersed in a 10% IPA aqueous solution (100 mL) under ultrasonic stirring for the photocatalytic H_2O_2 production under the irradiation of the light from a 300 W Xe lamp. The H_2O_2-production curves for K_X-CNs in comparison with the pristine CN are shown in Figure 5.8a, revealing that K^+-intercalation markedly accelerates the H_2O_2 production and the K_1-CN exhibits the highest H_2O_2 production. In Figure 5.8b, the H_2O_2-production rates are compared under different wavelength lights for K_X-CNs, with that for the K_1-CN catalyst reaching 11.2 mmol/g·h rate under the 350–780 nm light. The light with $\lambda = 350–780$ nm is much more effective to enhance H_2O_2-production rate than that with $\lambda = 400–780$ nm. In addition, the K_X-CNs can also decompose the H_2O_2 in the aqueous solution, as shown in a time dependence of H_2O_2 relative concentration in Figure 5.8c, $K_{0.5}$-CN exhibiting the fastest H_2O_2 decomposition. K^+-doping into *g*-CN was reported to promote the H_2O_2-production rate [12]. 10 wt% K^+ doping resulted in the increase in H_2O_2-production rate to 200 μmol/g·h, being 2.5 times the enhancement of the pristine *g*-CN. K-intercalated triazine (K-TCN) and tri-*s*-triazine units (K-TTCN) enhance H_2O_2-production performances in a water containing only 0.5 vol% IPA, being 11–32 times that of the *b*-CN [13].

Co-doping of K and P into *g*-CN was performed using a eutectic molten salt KCl/LiCl medium [14]. A mixture of melamine (5 g) and KCl/LiCl (0.408/0.592 by mole) with a mass ratio of melamine/molten salt (1/1, 1/3, and 1/5) was calcined at 550°C for 4 h to obtain K-doped CNs and coded as K_X-CN (K_1-CN, K_3-CN, and K_5-CN, respectively). To synthesize K- and P-co-doped CN, melamine (5 g) was mixed with $NH_4H_2PO_4$ (0.02, 0.05, and 0.1 g) and eutectic mixed salt KCl/LiCl (15 g) and then calcined at 550°C for 4 h. The resultant co-doped CNs were coded as $P_{0.02}K_3$-CN,

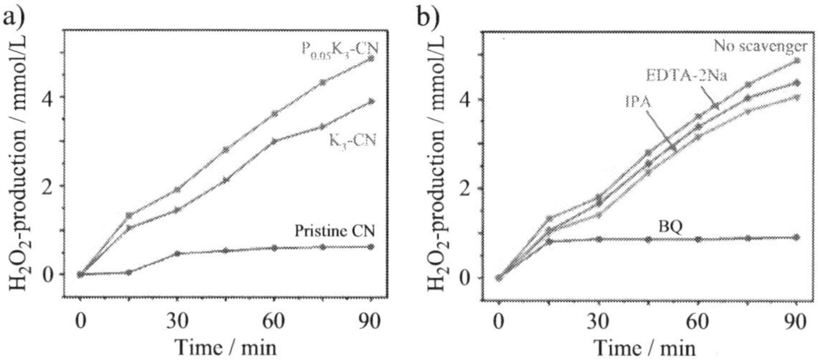

FIGURE 5.9 (a) H_2O_2-production kinetics for $P_{0.05}K_3$-CN, K_3-CN, and the pristine CN and (b) effects of scavengers on H_2O_2 production [14].

$P_{0.05}K_3$-CN, and $P_{0.1}K_3$-CN, respectively. H_2O_2-production performances were measured by dispersing 40 mg catalyst in a 10 vol% ethanol aqueous solution under bubbling of high-purity O_2 with a simulated solar light (100 mW/cm) from a 300 W Xe-lamp for 90 min. The H_2O_2 production over K_3CN was measured to be 3.82 mmol/L, and that over the co-doped composite $P_{0.05}K_3$-CN exhibited higher production of 4.97 mmol/L, while the H_2O_2 production over the pristine CN after 90 min irradiation is only 0.56 mmol/L, as shown in Figure 5.9a. The H_2O_2-production rate over these three catalysts is calculated to be 95.5, 124.3, and 14 mmol/g·h, respectively, revealing the importance of co-doping with K and P to obtain an enhanced photocatalytic activity of *g*-CN. In Figure 5.9b, the effects of the addition of different scavengers on H_2O_2 production over the $P_{0.05}$K-CN are shown, revealing that superoxide radicals ($*O_2^-$) play an important role in H_2O_2 production.

Co-doping of K^+ and cyano groups (–C≡N, Cy) into ultrathin *g*-CN (*ut*-CN) was achieved by an improved molten salt method to get the composite K_XCy-*ut*-CN [15]. A mixture of thiourea (2 g), NH_4Cl (4 g) with KCl (X g) (X=0.00, 0.03, 0.05, 0.07, and 0.09 g) was prepared via grinding and then heated at 520°C for 4 h in a closed crucible, and the resultant catalysts are coded as K_XCy-*ut*-CN. For comparison, K^+-doped CN was prepared via the same procedure without the addition of NH_4Cl (K-CN). H_2O_2 production was measured by dispersing 30 mg catalyst powder in 30 mL deionized water (containing 10% ethanol). In Figure 5.10a–c, H_2O_2-production kinetics are shown by focusing on the comparison of the resultant composites K_XCy-*ut*-CNs, the effect of pH of the reaction solutions on $K_{0.09}$Cy-*ut*-CN and the long-term stability of H_2O_2-production reaction over the $K_{0.00}$Cy-*ut*-CN and $K_{0.09}$Cy-*ut*-CN composites, respectively. With increasing KCl dosage, the yield of H_2O_2 production after 60 min irradiation was enhanced to 781.39 μmol from 91.08 μmol yield for the pristine CN (no KCy-co-dosage). The H_2O_2-production reaction continued steadily with increasing irradiation time up to 180 min.

Br and K-co-doped and cyano-group-rich CN (BrKCy-CN) was synthesized via calcining a molten mixture of urea and 6-bromopyridine-3-carbaldehyde (BrPyCHO), followed by re-calcination with potassium thiocyanate (KSCN) at 400°C or 2 h [16]. H_2O_2-production performances were measured by dispersing 10 mg catalyst in a 50 mL 10 vol% ethanol aqueous solution under the irradiation of visible light for

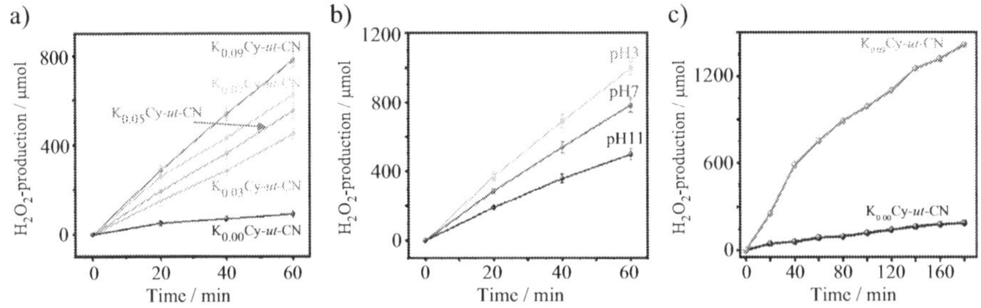

FIGURE 5.10 (a) H_2O_2-production kinetics for the resultant composite K_XCy-*ut*-CNs, (b) effect of pH of the reaction solution on K_9Cy-*ut*-CN, and (c) long-term stability of H_2O_2-production reaction over the $K_{0.00}$Cy-*ut*-CN and $K_{0.09}$Cy-*ut*-CN [15].

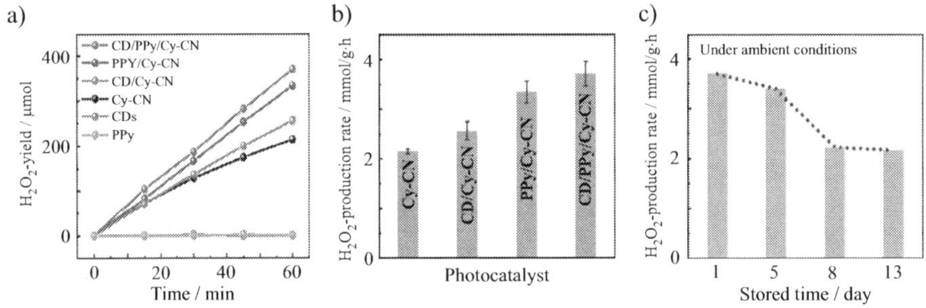

FIGURE 5.11 H_2O_2-production performances of the CD/PPy/Cy-CN in comparison with other related catalysts, PPy/Cy-CN, CD/Cy-CN, Cy-CN, CDs, and PPy: (a) kinetics, (b) rate of H_2O_2 production, and (c) storage-ability of the CD/PPy/Cy-CN [18].

30 min. The BrKCy-CN exhibited a high H_2O_2-production rate of 9.127 mmol/g·h. K-doped and surface hydroxy-group-rich CN was prepared by mixing melamine/cyanuric-acid-derived *g*-CN (0.5476 g) with 3.75 g KCl as well as different amounts of NaOH (0.0375, 0.05, and 0.0625 g), and being calcined and washed to remove the residual Na and Cl. The resultant catalysts as coded as $(OH)_3$K-CN, $(OH)_5$K-CN, and $(OH)_7$K-CN, respectively [17]. The $(OH)_5$K-CN catalyst exhibited the highest H_2O_2-production rate of 91.36 μmol/g·h in pure water under visible light irradiation.

First, a cyano-modified CN (Cy-CN) was prepared by heating a mixed solution of urea (10 g), KCl (10 g), and KOH (0.5 g) at 500°C for 4 h in air to get cyano-modified CN [18]. A cyano and carbon dots/polypyrrole (CDs/PPy) modified CN was prepared. Next, the cyano-groups and CDs/polypyrrole co-modified CN (CD/PPy/Cy-CN) was fabricated by in situ polymerization of pyrrole (PPy) on Cy-CN in the presence of CDs. The CDs were synthesized from a coal pitch. H_2O_2-production kinetics and H_2O_2-production rate over the resultant catalysts are compared in Figure 5.11a and b, respectively, revealing that the loading of PPy, CDs, and consequent loading of these two onto CNs are effective to enhance the photocatalytic activity for H_2O_2 production. The resultant CD/PPy/Cy-CN exhibited a high activity, the H_2O_2-production rate reaching 3.70 mmol/g·h. However, the H_2O_2-production rate over the CD/PPy/

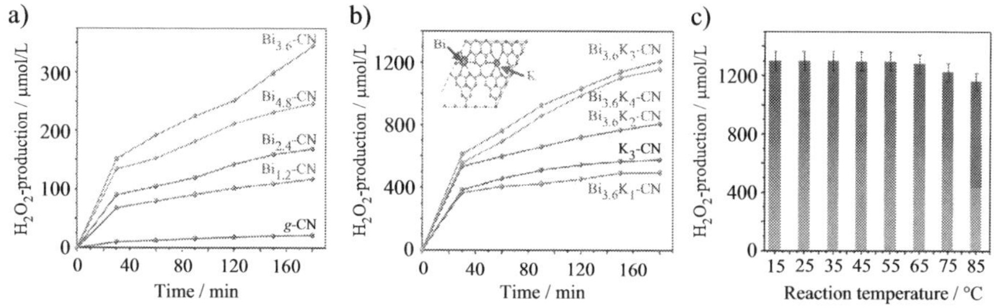

FIGURE 5.12 H_2O_2-production performances of the Bi and K co-doped CNs: (a) H_2O_2-production curves for the Bi-doped CNs, (b) Bi and K co-doped CNs with an illustration of Bi and K configurations (inset), and (c) temperature effect on the $Bi_{3.6}K_3CN$ [21].

Cy-CN showed a significant decrease after its storage for 10 days under ambient conditions, as shown in Figure 5.11c.

It was predicted by multiscale computational simulations that a co-intercalation of K^+ and I^- into a stacked *p*-CN can improve carrier separation and transfer [19]. As a consequence, the H_2O_2 production over the K^+ and I^- co-doped CN is efficiently enhanced, of which the rate was predicted to be 13.1 mmol/g·h. However, the paper worked on the synthesis of K and I co-doped *g*-CN reported its poor activity [20]. K^+ and I^- co-doped *g*-CN (IK-CN) was prepared by calcining a mixture of melamine (2.5 g) with 2.5 mmol KCl and KI at 550°C for 4 h, of which XRD analysis showed no change in d_{002}, indicating that there was no intercalation of both K and I ions [18].

Co-doped *g*-CN of Bi (1.2%, 2.4%, 3.6%, 4.8%) with K promoted the enhancement in H_2O_2 production [21]. H_2O_2-production curves are compared on Bi-doped CNs (Bi_X-CN, where *X* corresponds to Bi dosage) and the catalysts prepared by K-doping into $Bi_{3.6}CN$, $Bi_{3.6}K_Y$-CNs, where *Y* corresponds to K dosage (1–4) in Figure 5.12a and b, respectively. Both Bi-doping and K-doping can enhance the H_2O_2 production, while the enhancement became marked by co-doping of Bi and K into *g*-CN (Bi_XK_Y-CN, where *X* and *Y* correspond to the respective dosage), becoming more marked than the doping of either Bi or K. With increasing Bi-dosage up to 3.6%, H_2O_2 production enhanced (Figure 5.12a). The additional K-doping into $Bi_{3.6}$-CN resulted in further enhancement in H_2O_2 production, giving the H_2O_2 production of 1,208 μmol/L after 3 h (Figure 5.12b). The doped Bi is supposed to be kept in an intrinsic cavity coordinated by 6N V_{6N}, as the doped K does, as illustrated in Figure 5.12b as an inset. The photocatalytic activity for H_2O_2 production of the $Bi_{3.6}K_3$-CN was kept up to 55°C and gradually decreased above 55°C, as shown in Figure 5.12c.

O-doped *g*-CN with C-vacancies (V_{C3N}) was synthesized from supramolecular aggregates obtained from melamine with cyanuric acid mixed with oxalic acid by calcining in a lidded alumina crucible with microwave irradiation at 1,200 W power for 7 min [22]. The products were coded as V_{C3N}-O_X-CNs, where *X* corresponds to the amount of oxalic acid used (0.2, 0.4, and 0.6 g). H_2O_2-production performances were studied by dispersing 50 mg catalyst in a 50 mL water containing 10 vol% IPA under irradiation of the simulated solar light for 2 h. The resultant V_{C3N}-$O_{0.4}$-CN exhibited the highest H_2O_2 production of 2.01 mmol/h, being much higher than the pristine CN. The H_2O_2-production rate was calculated to be 40.16 mmol/g·h.

The introduction of N-vacancies into the *g*-CN framework was shown to be efficient to enhance H_2O_2-production performance [23]. To a DMF solution of Pluronic F-127, the solutions of melamine and cyanuric acid were added under stirring to obtain a white precipitate, which was calcined in a crucible at 550°C for different times (1, 3, 5, and 10 h). The products contained different amounts of N-vacancies (V_{N2C}) and coded as V_N-CN-*X*, where *X* corresponds to calcination time (1–10 h). H_2O_2-production rate depended strongly on calcination time *X*, V_N-CN-3 exhibiting the highest rate as 1,768 mmol/g·h.

By using the eutectic molten salt LiCl/KCl (EMS) with a melamine/EMS mass ratio of 1/2, a molecular heterostructure of heptazine and triazine motifs was obtained (coded as CN-2) [24]. This CN-2 exhibited a markedly enhanced H_2O_2-production rate in comparison with the PTI synthesized by conventional thermal polymerization by using an EMS mixing ratio of 1/10. The H_2O_2-production rate for CN-2 was determined to be 2,285 mmol/L·h, being about 143 times of the PTI (16 μmol/L·h). When the melamine/EMS mixing ratio changed to 1/1 and 1/5, the resultant photocatalysts exhibited a notably enhanced H_2O_2-production rate than PTI, but lower than CN-2.

A well-defined C_5N_2 compound with a conjugated C=N linkage was synthesized from a methanol solution of melamine with *p*-thioaldehyde in a 0.4/0.65 mass ratio by hydrothermal treatment at 180°C for 0.5–5 days [25]. The molecular structure of C_5N_2 is shown as an inset of Figure 5.13a. Its photocatalytic H_2O_2-production performances were investigated in a 150 mL pure water by dispersing 1.5 g of catalyst under the visible light irradiation from a Xe lamp with $\lambda > 400$ nm and 100 mW/cm^2 power. The results are shown in Figure 5.13a and b. The C_5N_2 catalyst exhibited a high H_2O_2-production rate of 1.55 mmol/L·h, being much higher than that of *g*-CN (5.2 μmol/L·h), as shown in Figure 5.13a. The H_2O_2-production rate of C_5N_2 was maintained after several repeated cycles (Figure 5.13b), suggesting its excellent stability.

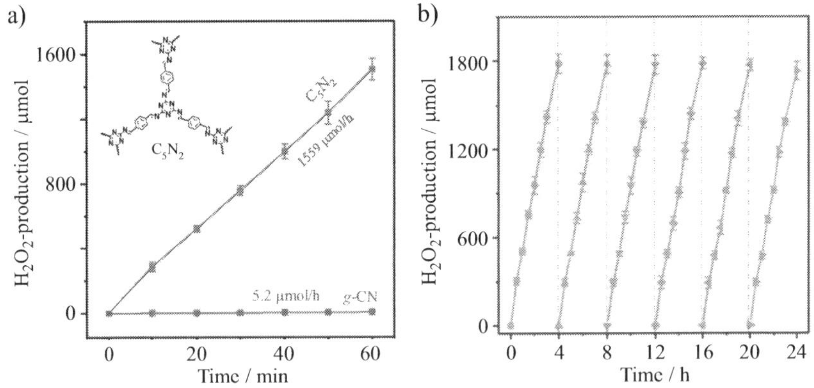

FIGURE 5.13 H_2O_2-production performances of C_5N_2 catalyst: (a) time changes of H_2O_2 production in comparison with the pristine *g*-CN and molecular structure of C_5N_2 as an inset and (b) cycle performance [25].

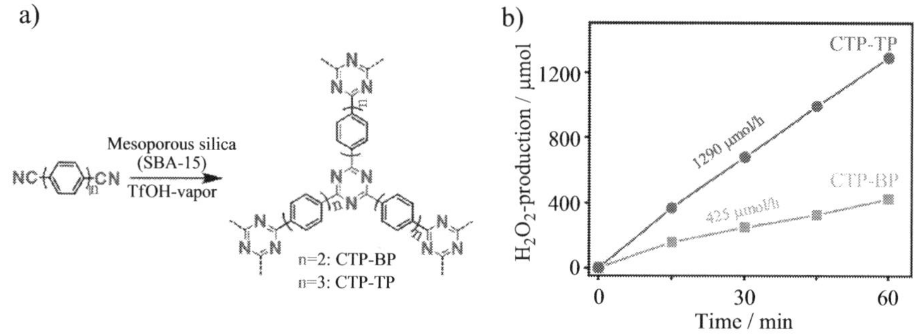

FIGURE 5.14 Covalent triazine polymers CTPs (CTP-BP and CTP-TP): (a) synthesis procedure and (b) H_2O_2 production in pure water under the irradiation of 10 W blue LED [26].

The covalent triazine polymers (CTPs) containing biphenyl (CTP-BP) and terphenyl (CTP-TP) units were prepared via cyano polymerization in the presence of trifluoromethyl sulfonic acid (TfOH) vapor using mesoporous silica (SBA-15) as the carrier [26]. The molecular structure of CTP-BP and CTP-TP is shown in Figure 5.14a. The H_2O_2-production performances of the resultant CTPs (CTP-BP and CTP-TP) were investigated by dispersing 6 mg catalyst powders in pure water under the irradiation of visible light by a 10 W blue LED. The results are shown in Figure 5.14b, revealing an excellent photocatalytic H_2O_2-production performance of the CTP-TP, its production rate reaching 1.29 mmol/h. H_2O_2-production rate of 2.013 mm/h could be achieved with the CTP-TP under Xe lamp irradiation ($\lambda > 420$ nm, 100 mW/cm^2).

5.2 *g*-C_3N_4-BASED COMPOSITES

The composites of N-rich CDs-loaded CN nanosheets having 0D/2D heterojunctions were prepared via hydrothermal calcination, which were coded as CD_X/CN, where X corresponds to the quantity of CDs loaded (0, 5, 10, 15, and 20 mg) [27]. The CDs were prepared by hydrothermal-calcination of an equimolar mixture of 1,8-diaminonaphthalene and DCDA at 180°C for 12 h. The mixture of melamine and CDs was heated at 550°C for 2 h to get CDs-loaded CN, which was heated further at 640°C for 4 h to obtain the composite CD_X/CNs. On the resultant composites, the H_2O_2-production performance was studied in O_2-saturated solutions employing IPA as a hole sacrificial agent under visible light ($\lambda \geq 420$ nm) for 3.5 h. The composite CD_{10}/CN exhibited a high H_2O_2 production of 1.48 mmol/L, approximately 10 times that of the pristine CN (i.e., CD_0/CN), while other composites CD_5/CN, CD_{15}/CN, and CD_{20}/CN exhibited comparable to the pristine CN.

A composite UiO/IK-CN composed of I^- and K^+ co-doped CN (IK-CN) with NH_2-UiO-66 (MOF) exhibited a high H_2O_2-production performance [28]. H_2O_2-production yield and rate over UiO/IK-CN composite photocatalyst is compared with those of the related catalysts, i.e., the pristine CN, NH_2-UiO-66 and IK-CN, under visible light irradiation ($\lambda > 400$ nm) in Figure 5.15a and b, respectively. The 10 wt% loading of MOF was determined to be the optimal for H_2O_2 production. After 90 min irradiation, the H_2O_2 yield of IK-CN (494.5 µmol) was 20 times that of

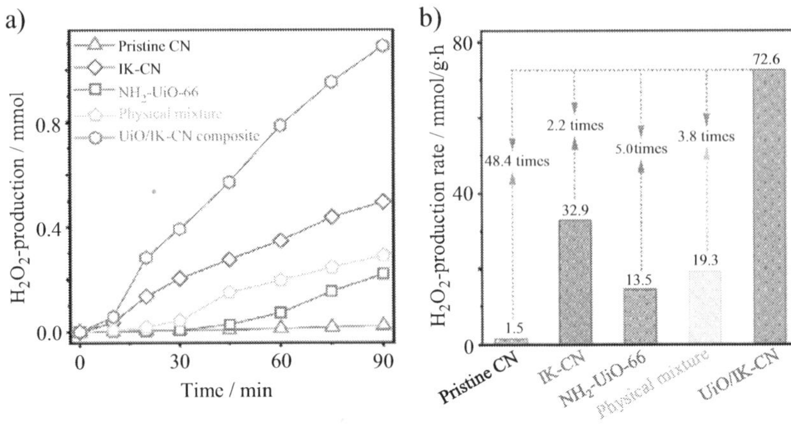

FIGURE 5.15 H_2O_2-production performances of UiO-66/IK-CN composite in comparison with the pristine CN, IK-CN, NH_2-UiO-66 (MOF) and a physical mixture of IK-CN with a MOF (NH_2-UiO-66): (a) H_2O_2-production yield and (b) production rate [28].

the pristine CN (22.5 μmol), revealing the effectiveness of I^- and K^+ co-doping. The coupling of NH_2-UiO-66 with IK-CN further improved the photocatalytic performance significantly. The UiO/IK-CN composite showed the highest H_2O_2-production rate (72.6 mmol/g·h) in comparison with those of all related catalysts, 48.4, 2.2, 5.0, and 3.8 times higher than that of the pristine CN, IK-CN, NH_2-UiO-66 (MOF), and a physical mixture of IK-CN with MOF, as shown in Figure 5.15b.

IK-CN was loaded onto CdSe-diethylenetriamine (CdSe-D) to enhance the H_2O_2-production rate [20]. The composite $(KI\text{-}CN)_X$/CdSe-D was prepared by adding IK-CN into a suspension of CdSe-D composite in different mass ratios X to CdSe-D (X = 30%, 40%, 50%, 60%). The composite CdSe-D was prepared by microwave hydrothermal treatment of the solution mixed with $CdCl_2$ and Na_2SeO_3 (in 1/1 by mole) in diethylenetriamine and $N_2H_4 \cdot H_2O$ (2/1 by volume) solution at 140°C for 20 min. For comparison, K-doped *g*-CN (K-CN) and I-doped CN (I-CN) were also prepared. The H_2O_2 production was measured by dispersing 10 mg of catalysts in 50 mL O_2-saturated deionized water under a continuous flow of O_2 and an irradiation of the simulated solar light from a 300 W Xe lamp. H_2O_2-production kinetics and rate are shown in comparison with the pristine CdSe-D and K^+I^--CN in Figure 5.16a and b, respectively. The H_2O_2 yield of CdSe-D reached 204.92 μmol after 1 h irradiation, whereas K^+, I^--CN exhibited negligible H_2O_2 under the same conditions. The compositing of K^+, I^--CN with CdSe-D markedly enhanced the yield and the rate of H_2O_2 production. The $(K^+/I^-\text{-CN})_{40}$/CdSe-D composite exhibited the highest H_2O_2-yield of 448.6 μmol after 1 h, and the production rate was calculated to reach 2.24 μmol/h·g.

The self-composite AOP-CN, which was composed of a ternary homojunction between amino-rich CN (A-CN), oxidized CN (O-CN), and phosphorylated CN (P-CN), delivered a high performance for H_2O_2 production [29]. The H_2O_2-production kinetics in pure water (without sacrificial agent) and with IPA as a sacrificial agent by dispersing 300 mg catalyst are shown in Figure 5.17a and b, respectively. The composite AOP-CN hydrogel exhibited the highest H_2O_2-production performances in both

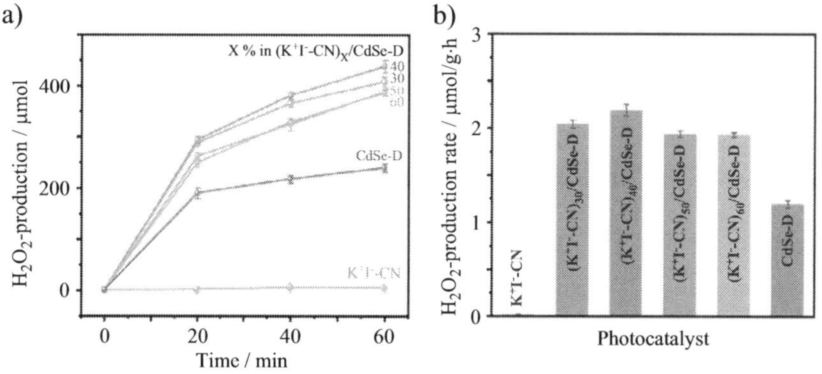

FIGURE 5.16 H_2O_2-production performances of (K^+, I^--CN)$_X$/CdSe-D composites in comparison with the K^+,I^--CN and CdSe-D: (a) H_2O_2-production kinetics and (b) H_2O_2-production rate [20].

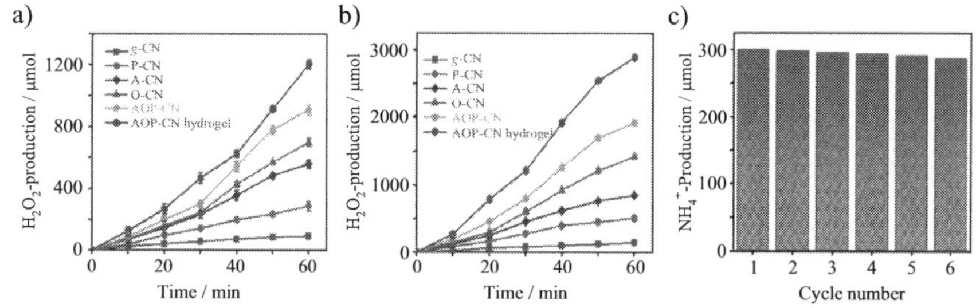

FIGURE 5.17 Photocatalytic H_2O_2 production over the AOP-CN composite with ternary homojunctions in comparison with the related catalysts: (a) without sacrificial agent, (b) using isopropanol (sacrificial agent), and (c) reusability [29].

cases (with or without sacrificial agent), reaching 2,890–1,204 μmol in 60 min of visible light irradiation. The hydrogelation of the composite was effective in enhancing H_2O_2 production. The kinetic studies confirmed that the homojunction construction in the composite reduced the H_2O_2 decomposition, in addition to an improvement in the H_2O_2 production. The composite hydrogel exhibited a high reusability, six cycles showed a decrease by only 56 μmol of H_2O_2 production, as shown in Figure 5.17c.

A ternary composite Bi/$Bi_4Ti_3O_{12}$/*g*-CN (Bi/BTO/CN_X) with van der Waals heterojunction was fabricated by a facile hydrothermal method [30]. $Bi(NO_3)_3$·(0.5 g) and different amounts of melamine-derived *g*-CN (X = 10, 20, 30, and 40 mg) were dispersed in 3 M NaOH solution, which was ultrasonically stirred before undergoing hydrothermal treatment at 200°C for 48 h. The product Bi/BTO/CN_X composites were labeled as Bi/BTO/CN_{10}–Bi/BTO/CN_{40}, respectively. The BTO is known to be formed as nanosheets with piezo-catalytic activity [31]. The H_2O_2-production performances of these composites were studied by dispersing 100 mg of composite in 100 mL of mixed solution of alcohol/water (1/9) and measuring the concentration of H_2O_2 during the piezo-catalyzed reaction at room temperature. As shown in

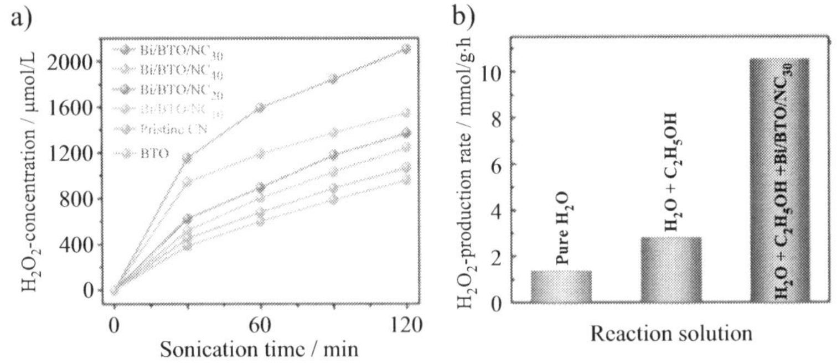

FIGURE 5.18 H_2O_2-generation performance with piezo-catalytic activity of the composites Bi/BTO/CN_X in comparison with the pristine CN and BTO: (a) H_2O_2-generation kinetics with sonication time on the composites, and (b) the effect of reaction solution for the composite Bi/BTO/CN_{30} [30].

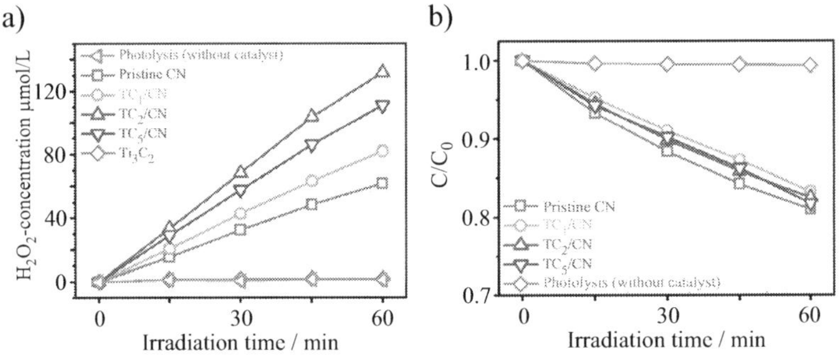

FIGURE 5.19 $(Ti_3C_2)_X$/CN composites: (a) H_2O_2-production kinetics in comparison with photolysis, the pristine CN and Ti_3C_2 and (b) H_2O_2-decomposition kinetics with an initial H_2O_2 concentration of 1 mmol/L [32].

Figure 5.18a, a progressive increase in the H_2O_2-concentration was observed with increasing ultrasonication time. The H_2O_2-generation rates for the composites Bi/BTO/CN_X were higher than those of the pristine CN and BTO. The composite Bi/BTO/CN_{30} reveals the highest performance, giving the rate of 10.5 mmol/g, larger than those for BTO (4.81 mmol/g·h) and CN (5.33 mmol/g·h). In Figure 5.18b, an effect of reaction solution is shown, the absence of alcohol or catalyst resulting in a low H_2O_2-production rate.

Ti_3C_2 (TC, MXene) nanosheets were loaded onto porous *g*-CN, which was prepared by controlling the mass ratio of TC to *g*-CN to 1%, 2%, and 5% [32]. The resultant composites are coded as TC_X/CN, where *X* corresponds to the mass ratio of TC loaded (1%, 2%, and 5%). H_2O_2 production was measured by dispersing 50 mg of catalyst powder in 50 mL of water containing 10 vol% IPA under the irradiation of light with a 300 W Xe lamp equipped with a 420 nm cut-off filter and then stirring. The H_2O_2-production kinetics on the resultant composites are shown in Figure 5.19a, revealing that the composite

TC_2/CN exhibited the highest H_2O_2 production, with the rate reaching 131.71 μmol/g·h. The H_2O_2 decomposition behavior over the photocatalysts was also studied by dispersing 50 mg of catalysts in 50 mL of 1 mmol/L H_2O_2 solution under the irradiation of visible light and continuous stirring for 1 h. As shown in Figure 5.19b, H_2O_2 decomposed about 17.5% on TC_2/CN after 1 h of visible light irradiation, the decomposition rate constant was calculated as $3.21 \times 10^{-3} min^{-1}$.

A composite of the Mo-doped WO_3 nanowires and the In-doped CN nanosheets (Mo-WO_3/In-CN) exhibited enhanced H_2O_2-production performances [33]. In-CN was prepared by calcining a mixture of urea (10 g) with $InCl_2$ (5 mg) at 550°C for 3 h. The Mo-doped WO_3 was obtained by hydrothermal treatment of an aqueous solution of $Na_2WO_4{\cdot}2H_2O$ (1.0 g) with $(NH_4)_6Mo_7O_{24}{\cdot}4H_2O$ (0.1 g) at 180°C for 12 h. The powder mixtures of 100 mg In-CN with various amounts of Mo-WO_3 (10–40 mg) were heated at 520°C for 2 h in Ar to synthesize the composite (Mo-$WO_3)_X$/In-CN, where X corresponds to the mass ratio to In-CN (0.1, 0.2, 0.3, and 0.4 mg). The H_2O_2-yield by the (Mo-$WO_3)_{0.2}$/In-CN composite achieves 6,165.8 μmol/g after 2 h irradiation (3,082.9 μmol/g·h) in pure water with O_2-bubbling, which is 8.6 times that of In-CN and 33.4 times that of Mo-WO_3.

A $TiO_{2-x}/g\text{-}C_3N_{4-x}$ was prepared via introducing O and N vacancies into the anatase-type TiO_2 sheets and the $g\text{-}C_3N_4$ matrices through a one-step reduction process [34]. An *n*-propanol solution of the urea-derived *g*-CN (1.1 g) and the tetra-n-butyl orthostates (4 mL) was slowly added to 1.6 mL HF, followed by hydrothermal treatment at 180°C for 24 h. The resultant precipitates were further heated at 300°C for 9 h to obtain the TiO_2-loaded *g*-CN, which was reduced using $NaBH_4$ at 300°C for 2 h in Ar to synthesize the reduced composite $TiO_{2-x}/g\text{-}C_3N_{4-x}$, which contains O-vacancies in TiO_2 and N-vacancies in *g*-CN. The H_2O_2 production was measured on these catalysts in pure water under a simulated solar light at room temperature. The H_2O_2-production kinetics are shown in Figure 5.20a. H_2O_2-yield after 60 min irradiation over TiO_2 (anatase) and TiO_{2-x} are measured to be 39.42 and 91.12 μmol/L, respectively (2.31-fold increase), and that over $g\text{-}C_3N_4$ and $g\text{-}C_3N_{4-x}$ are 143.08, 193.37 μmol/L, respectively (1.35-fold increase). These results demonstrate the effectiveness of the vacancy-introduction in respective photocatalysts, TiO_2 and $g\text{-}C_3N_4$.

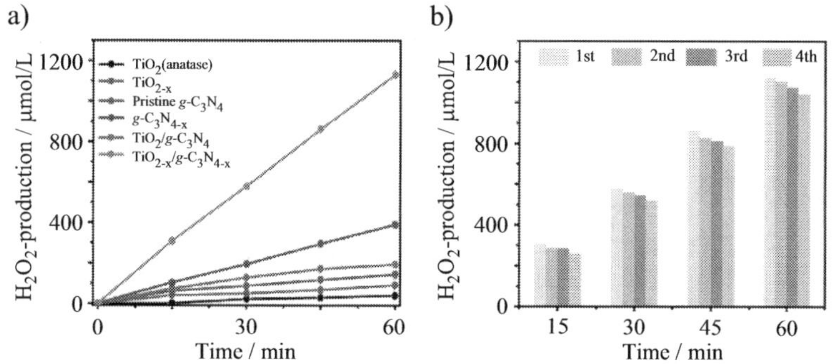

FIGURE 5.20 H_2O_2-production performances of $TiO_{2-x}/g\text{-}C_3N_{4-x}$ in comparison with the related catalysts: (a) kinetics and (b) stability [34].

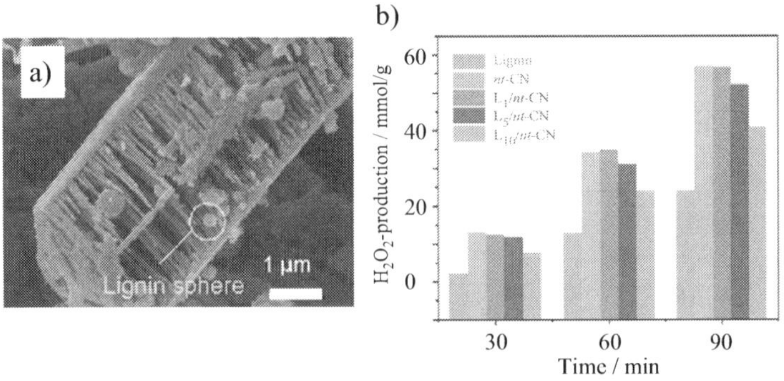

FIGURE 5.21 Lignin nanosphere-loaded tubular *g*-CN (L_X/*nt*-CN): (a) SEM image of L_{10}/*nt*-CN cross-section and (b) H_2O_2-yield over the composites in comparison with lignin itself [35].

Furthermore, H_2O_2-yield over TiO_2/*g*-C_3N_4 and TiO_{2-x}/*g*-C_3N_{4-x} are 398.4 and 1,131.59 μmol/L, respectively (2.84-fold increase), suggesting synergetic enhancement through compositing O-defective photocatalyst (TiO_2) with N-defective photocatalyst (*g*-C_3N_4). The H_2O_2 production of TiO_{2-x}/*g*-C_3N_{4-x} composite photocatalyst remains essentially unchanged after four cycles, as shown in Figure 5.20b, confirming the stability of its photocatalytic performance.

Loading lignin nanospheres onto tubular *g*-CN (*nt*-CN) enhanced H_2O_2 production [35]. *nt*-CNs were synthesized by hydrothermal heating of a mixture of melamine (1 g) with phosphorous acid (1 g) in 70 mL deionized water at 180°C for 10 h and then washing to remove phosphorous completely, followed by heating at 500°C for 4 h. The *nt*-CN (1 g) was added into aqueous dispersion lignin nanospheres (10 mL) and freeze-dried to obtain lignin-loaded *nt*-CNs (L_X/*nt*-CN, where *X* corresponds to the quantity of lignin nanospheres used). In Figure 5.21a, an SEM image of a cross-section of L_{10}/*nt*-CN is shown by highlighting the lignin nanosphere. H_2O_2-yields after different times of visible light irradiation are compared on the resultant composites L_X/*nt*-CNs, including lignin itself in Figure 5.21b. H_2O_2-yields after 1 h irradiation were calculated as 22.8, 44.2, 43.7, 41.1, and 33.9 mmol/g for the lignin itself, *nt*-CN, L_1/*nt*-CN, L_5/*nt*-CN, and L_{10}/*nt*-CN, respectively. Compositing with lignin nanospheres did not give a marked benefit for H_2O_2 production. However, the compositing gave certain benefits for the photodegradation of BBBN; refer Chapter 7.

5.3 CONCLUDING REMARKS

H_2O_2-production performances over *g*-CN-related catalysts are summarized by focusing on H_2O_2-production rate together with characteristics of *g*-CN in Table 5.1. As pointed out in Table 3.5 in Chapter 3 on H_2-evolution, the crystallinity of the *g*-CN also seems to govern the H_2O_2-generation rate, because the *g*-CN regenerated using $ZnCl_2$ molten salt hydrates with exfoliation to nanobelts exhibited 2 orders of magnitude larger rate in comparison with the bulk *g*-CN synthesized from different CN-precursors [7]. In addition, doping of B into CN (B-CN) is effective in enhancing

TABLE 5.1
H_2O_2-Production Performances of *g*-CN-Based Photocatalysts

Photo-Catalyst	Synthesis and Modification of *g*-CN: Conditions		Synthesis and Modification of *g*-CN: Character	H_2O_2 Production: Co-Catalyst, etc.	H_2O_2 Production: Rate (mmol/h·g)	References
			g-C_3N_4 and Modified g-C_3N_4			
g-CN	Cyanamide			Water + ethanol 25°C	0.125	[3]
g-CN	Urea + KCl or KOH		Abundant active surfaces	Pure water	0.22	[4]
a-CN	Melamine + cyanuric acid Supramolecular aggregates	H_2-plasma	Amorphous CN (*a*-CN)	Water + isopropanol	1.656	[5]
g-CN		550°C	Bulk CN		0.507	
b-CN	Melamine	550°C	Pristine CN		0.824	
fl-CN-530	Exfoliation using nitrate, at 530°C		Few-layered sheets	Water + ethanol, O_2-bubbling	0.952	[6]
BP-CN	Black phosphorus Intercalated		Liquid exfoliation	Water + isopropanol	0.54	[8]
B-CN-350	B-doping $NaBH_4$ at 350°C				51.008	[9]
O-CN-500	DCDA + 1 mol% ammonium para-tungstate at 500°C		O-enriched *g*-CN (O-CN)	Water + 10% isopropanol	58.4	[10]
K_1-CN	Urea + KCl		K^+-intercalated CN	Water + 10% isopropanol	11.2	[11]
$P_{0.05}K$-CN	Melamine + KCl/LiCl + $NH_4H_2PO_4$		P-, K-co-doping	10 vol% ethanol + water, O_2-bubbling	124.300	[14]
K_9Cy-*ut*-CN	Thiourea + NH_4Cl + K		K-doping and cyano group modified		26.046[a]	[15]
CD/PPy/Cy-CN	Carbon dots + PPy-loaded on Cy-CN		Cyano-group-rich CN	Pure water	3.70	[18]

(Continued)

TABLE 5.1 (*Continued*)
H_2O_2-Production Performances of *g*-CN-Based Photocatalysts

	Synthesis and Modification of *g*-CN		H_2O_2 Production		
Photo-Catalyst	Conditions	Character	Co-Catalyst, etc.	Rate (mmol/h·g)	References
IK-CN		I^- and K^+-co-doping		13.1	[19]
V_{C3N}-$O_{0.4}$-CN	O-doping C-vacancy (V_{C3N}) Microwave-heating	Supramolecular aggregates	Water with 10 vol% IPA	40.16	[22]
V_N-CN-3	Melamine + cyanuric acid with F-127	Containing N-vacancy (V_{N2C})	Water + 20 vol% methanol	1.768	[23]
		***g*-C_3N_4-Based Composites**			
UiO/IK-CN	10 wt% NH_2-UiO-66 (MOF)-loading	I^- and K^+ co-doped CN (IK-CN)		72.6	[28]
$(IK\text{-}CN)_{40}$/CdSe-D	IK-CN-loaded onto CdSe-D		O_2-saturated water with O_2 flow	2.24	[20]
AOP-CN	Self-composite of amino-rich CN (A-CN), oxidized CN (O-CN) and phosphorylated CN (P-CN)		Without IPA	4.01[a]	[29]
			With IPA	9.63[a]	
Bi/BTO/CN_{30}	Loading Bi-nanoparticles and $Bi_4Ti_3O_{12}$ nanosheets-loading		Alcohol/water (1/9) mixture	10.5 (Piezo-catalytic)	[30]
TC_2/CN	2%Ti_3C_2 (TC) nanosheets-loading		Water with 10 vol% IPA	0.132	[32]
$(Mo\text{-}WO_3)_{0.2}$/In-CN	Mo-doped WO_3-loading	In-doped CN (In-CN)	Pure water with O_2-bubbling	3.08	[33]
TiO_{2-x}/*g*-C_3N_{4-x}	O-deficient TiO_2-loading	N-deficient *g*-CN	Pure water	1.31	[34]
ut-CN		CN nanotube (*nt*-CN)	N_2-saturation	44.2	[35]
L_1/*ut*-CN	Lignin nanospheres-loading			43.7	

[a] Recalculated to the common unit (mmol/g·h) by taking into account the amount of catalyst used.

H_2O_2 production. B-CN obtained via a 350°C -calcination resulted in the rate of 51.008 mmol/g·h, while via a 450°C -calcination gave 45.720 mmol/g·h [9].

A ternary composite $Bi/Bi_4Ti_3O_{12}/g$-CN ($Bi/BTO/NC_{30}$) with 2D/2D heterojunctions exhibited a certain H_2O_2-production rate only when it was subjected to ultrasonic vibration (400 W, 40 kHz), which appears to be primarily due to the piezo-catalytic effect of Bi/BTO. Sonication under visible light irradiation may allow the photocatalytic activity of *g*-CN to collaborate with piezo-catalytic activity to enhance the whole H_2O_2-production rate.

REFERENCES

1. Wang Y, Waterhouse GIN, Shang L, et al., Electrocatalytic oxygen reduction to hydrogen peroxide: From homogeneous to heterogeneous electrocatalysis. *Adv Energy Mater* 2020; 11: 1802525.
2. Yang Z, Li L, Zeng S, Cui J, Wang K, Hu C, Zhao Y, Nanoarchitecture manipulation by polycondensation on KCl crystals toward crystalline lamellar carbon nitride for efficient H_2O_2 photoproduction. *ACS Appl Mater Interfaces* 2023; 15: 8232–40.
3. Shiraishi Y, Kanazawa S, Sugano Y, et al., Highly selective production of hydrogen peroxide on graphitic carbon nitride (g-C_3N_4) photocatalyst activated by visible light. *ACS Catal* 2014; 4: 774–80.
4. Yue J, Yang H, Liu C, et al., In-plane electron transfer dominated carbon nitride for high-efficiency quasi-homogeneous photochemical synthesis of hydrogen peroxide. *Chem Eng J* 2024; 497: 154683.
5. Zheng Y, Luo Y, Ruan Q, et al., Plasma-induced hierarchical amorphous carbon nitride nanostructure with two N_2C-site vacancies for photocatalytic H_2O_2 Production. *Appl Catal B* 2022; 311: 121372.
6. Feng B, Liu Y, Wan K, et al., Tailored exfoliation of polymeric carbon nitride for photocatalytic H_2O_2 production and CH_4 valorization mediated by O_2 activation. *Angew Chem Inter Ed* 2024; 63(18): e202401884.
7. Shen D, Imbault AL, Balati G, et al., Dissolution of g-C_3N_4 using zinc chloride molten salt hydrates for nanobelt fabrication and photocatalytic H_2O_2 production. *Chem A Eur J* 2024; 30 (18): e202401847.
8. Zheng Y, Yu Z, Ou H, et al., Black phosphorus and polymeric carbon nitride heterostructure for photoinduced molecular oxygen activation. *Adv Funct Mater* 2018; 28: 1705407.
9. Ma L, Gao Y, Wei B, et al., Visible-light photocatalytic H_2O_2 production boosted by frustrated Lewis pairs in defected polymeric carbon nitride nanosheets. *ACS Catal* 2024; 14: 2775–86.
10. Wei Z, Liu M, Zhang Z, et al., Efficient visible-light-driven selective oxygen reduction to hydrogen peroxide by oxygen-enriched graphitic carbon nitride polymers. *Energy Environ Sci* 2018; 11: 2581–9.
11. Zhang Q, Wang B, Miao H, et al., Efficient photocatalytic H_2O_2 production over K^+-intercalated crystalline g-C_3N_4 with regulated oxygen reduction pathway. *Chem Eng J* 2024; 482: 148844.
12. Haripriya P, Anjana T, Sreelakshmi K, et al., Effect of alkali metal cation doping in graphitic carbon nitride towards photocatalytic generation of hydrogen peroxide under direct sunlight. *Catal Commun* 2024; 187: 106909.
13. Zhang J, Yu C, Lang J, et al., Modulation of Lewis acidic-basic sites for efficient photocatalytic H_2O_2 production over potassium intercalated tri-s-triazine materials. *Appl Catal B* 2020; 277: 119225.

14. Liang H, Zhao J, Wang A, et al., Synthesis of novel nanoflowers-like P, K co-doped graphitic carbon nitride for efficient H_2O_2 photoproduction. *J Colloid Interface Sci* 2024; 677(B): 729–39.
15. Huang X, Zhou C, Yang J, et al., Potassium/cyano group co-incorporation promotes 2e-ORR selectivity in porous ultrathin carbon nitride for photocatalytic H_2O_2 production. *Solid State Sci* 2024; 156: 107669.
16. Fang H, Zhang Y, Xu J, et al., Introducing Br, K, and cyano group into carbon nitride for efficient photocatalytic hydrogen peroxide production then in situ tetracycline mineralization. *J Colloid Interface Sci* 2024; 667: 433–40.
17. Wu J, Yu J, Fan F, et al., Constructing potassium and hydroxyl co-doped dual-dipole structures on highly active 3D g-C_3N_4 surfaces for highly boosting photocatalytic hydrogen peroxide production efficiency in pure water. *Green Chem Eng* 2024; 6(3): 388–97.
18. Zhang Y, Li X, Ding S, et al., Boosting photocatalytic H_2O_2 generation by constructing a cyano and carbon dots/polypyrrole co-modified carbon nitride heterostructure. *Adv Sustain Syst* 2024; 8(10): 2400261.
19. Liu LL, Chen F, Wu JH, et al., Synergy of crystallinity modulation and intercalation engineering in carbon nitride for boosted H_2O_2 photosynthesis. *Proc Natl Acad Sci USA* 2023; 120: 2215305120.
20. He H, Wang Z, Zhang J, et al., Interface chemical bond enhanced ions intercalated carbon nitride/CdSe-diethylenetriamine S-scheme heterojunction for photocatalytic H_2O_2 synthesis in pure water. *Adv Funct Mater* 2024; 34(24): 2315426.
21. Liang H, Wang A, Cheng R, et al., Bi, K co-doped graphitic phase carbon nitride for efficient photocatalytic H_2O_2 production. *Chem Eng J* 2024; 489: 151145.
22. Xie H, Zheng Y, Guo X, et al., Rapid microwave synthesis of mesoporous oxygen-doped g-C_3N_4 with carbon vacancies for efficient photocatalytic H_2O_2 production. *ACS Sustain Chem Eng* 2021; 9: 6788–98.
23. Zhao C, Shi C, Li, Q, et al., Nitrogen vacancy-rich porous carbon nitride nanosheets for efficient photocatalytic H_2O_2 production. *Mater Today Energy* 2022; 24: 100926.
24. Wang Y, Zhang H, Wei D, et al., Molten-salt-induced structural heterogeneity of carbon nitride for efficient photocatalytic H_2O_2 production. *Chem Commun* 2024; 60: 10732.
25. Ma J, Peng X, Zhou Z, et al., Extended conjugation tuning carbon nitride for non-sacrificial H_2O_2 photosynthesis and hypoxic tumor therapy. *Angew Chem Int Ed* 2022; 61(43): e202210856.
26. Zhang Q, Zhou J, Zhang H, et al., Terphenyl functionalized covalent triazine polymer as metal-free photocatalyst for superior hydrogen peroxide production. *Adv Funct Mater* 2024; 34(32): 2401579.
27. Guo H, Zhou L, Huang K, et al., Nitrogen-rich carbon dot-mediated $n\pi^*$ electronic transition in carbon nitride for superior photocatalytic hydrogen peroxide production. *Adv Funct Mater* 2024; 34(38): 2402650.
28. Lu X, Chen Z, Hu Z, et al., Boosted charge transfer for highly efficient photosynthesis of H_2O_2 over Z-scheme I^-/K^+ co-doped g-C_3N_4/metal–organic-frameworks in pure water under visible light. *Adv Energy Mater* 2024; 14(38): 2401873.
29. Balakrishnan A, Chinthala M, Kumar A, et al., Construction of dual Z-scheme ternary carbon nitride homojunction pectin microspheres as a multifunctional photocatalyst for tetracycline degradation, H_2O_2 production, and N_2 fixation. *Chem Eng J* 2024; 496: 153899.
30. Xu Q, Zhang Y, Lu D, et al., Bi nanoparticle/$Bi_4Ti_3O_{12}$ nanosheet/g-C_3N_4 nanowire heterojunction for the piezocatalytic H_2O_2 production. *ACS Appl Nano Mater* 2024; 7: 17379–90.
31. Wu J, Qin N, Lin E, et al., Synthesis of $Bi_4Ti_3O_{12}$ decussated nanoplates with enhanced piezocatalytic activity. *Nanoscale* 2019; 11(44): 21128–36.

32. Yang Y, Zeng Z, Zeng G, et al., Ti_3C_2 Mxene/porous g-C_3N_4 interfacial Schottky junction for boosting spatial charge separation in photocatalytic H_2O_2 production. *Appl Catal B* 2019; 258: 117956.
33. Zhang R, Wu J, Li K, et al., Construction of a 1D/2D S-scheme heterojunction of Mo-doped WO_3/In-doped carbon nitride for efficient hydrogen peroxide photosynthesis. *J Phys Chem C* 2024; 128: 5837–47.
34. Gao J, Yang S, Wei P, et al., Enhanced photocatalytic H_2O_2 production through synergistic double-vacancy engineering in Z-scheme TiO_{2-x}/g-C_3N_{4-x} heterojunction. *Sep Purif Technol* 2025; 354(1): 128685.
35. Zhu C, Xiao X, Wang X, et al., Lignin-modified graphitic carbon nitride nanotubes for photocatalytic H_2O_2 production and degradation of brilliant black BN. *Inter J Biol Macromol* 2024; 267(2): 131533.

6 Nitrogen Fixation by g-C_3N_4

Electrocatalytic nitrogen reduction reaction (NRR) under ambient conditions provides an intriguing picture for the conversion of N_2 into ammonia (NH_3). Since NH_3, the nitrogen source of manmade fertilizer, is one of the most fundamental synthetic chemicals for sustaining human life. Although N_2 from the atmosphere is inexhaustible, its chemical inertness makes the conversion of N_2 into NH_3 difficult. Industrially, the synthesis of NH_3 is often accomplished relying on an iron-based catalyst and harsh conditions (300–400°C and ca. 250 atm), known as the Haber-Bosch process. Such process accounts for 1.4% of all energy consumed by mankind, accompanied by 1.6% of total global CO_2 emissions.

Therefore, photocatalytic NRR remains a grand challenge in enabling effective N_2 activation on metal-free catalysts. The fundamental reaction for N_2 fixation for NH_4^+ formation with the protons from H_2O (Eq. 6.3) can be expressed by the following equations:

$$g-\mathrm{CN} + h\nu \rightarrow e^- + h^+ \tag{6.1}$$

$$3\mathrm{H_2O} + 6h^+ \rightarrow 2/3\mathrm{O_2} + 6\mathrm{H}^+ \tag{6.2}$$

$$\mathrm{N_2} + 6\mathrm{H}^+ + 6e^- \rightarrow 2\mathrm{NH_3} \tag{6.3}$$

6.1 g-C_3N_4 AND MODIFIED g-C_3N_4

The polymeric carbon nitride (p-CN) obtained by polycondensation of melamine at 550°C was further calcined at 620°C for 2, 4, and 6h in Ar flow to create N-vacancy (V_N) [1]. The resultant p-CNs are coded as V_N-CN-X (X represents the further calcination hours in Ar). Electrocatalytic N activation to form NH_3 on the resultant catalyst V_N-CN-4 was measured by a potentiostatic procedure using the catalyst cathode with 0.1M HCl solution electrolyte at room temperature and atmospheric pressure with a continuous N_2 stream to the cathode. The results are shown in Figure 6.1a–c. In Figure 6.1a, NH_3-yields of the resultant V_N-CN-Xs are compared to the pristine p-CN, revealing that the V_N-CN-4 exhibits the highest NH_3 yield, and the introduction of N-vacancies into p-CN is very effective to enhance the NH_3-yield to more than 10 times. In Figure 6.1b and c, the highest NH_3-yield of 8.09 μg/h·mg and Faradaic efficiency of 11.59% are obtained at the potential of −0.2V on the metal-free p-CN catalyst with N-vacancies. The experimental results demonstrated that a defect engineering strategy was useful to obtain enhanced NRR performance.

DOI: 10.1201/9781003589051-6

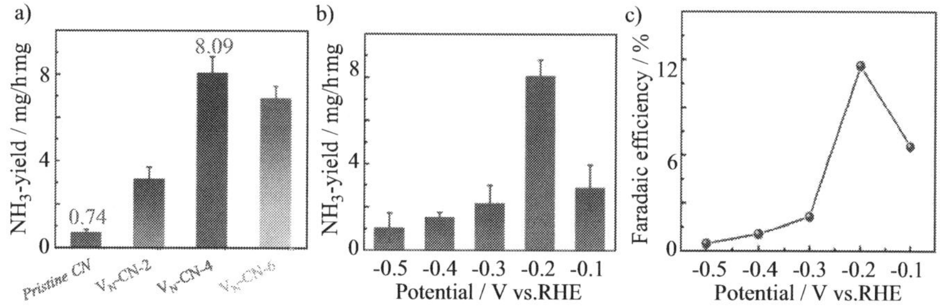

FIGURE 6.1 Electrocatalytic NRR of V_N-CN-X catalysts at ambient conditions: (a) average yield of NH_3 in comparison with the pristine p-CN, (b) yield of NH_3 against potential applied, and (c) Faradaic efficiency with potential on V_N-CN-4 [1].

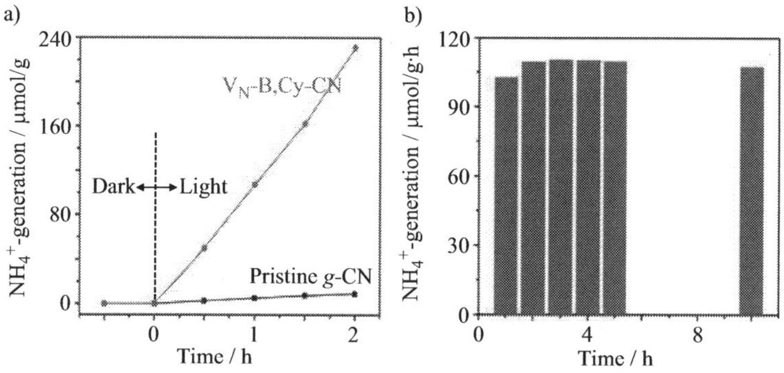

FIGURE 6.2 N_2-fixation performances of B and cyano-group (Cy) co-doped g-CN with nitrogen vacancies (V_N-B, Cy-CN) in comparison with the pristine g-CN: (a) NH_4^+-generation kinetics and (b) time change of the rate [2].

B and cyano-group (Cy) co-doped g-CN with N-vacancies (V_N-B, Cy-CN) was synthesized from a mixture of dicyandiamide (DCDA) and boric acid by calcining at 600°C for 4 h [2]. In Figure 6.2a and b, N_2-fixation performances are shown by NH_4^+-generation kinetics and the time change of its rate, respectively. Over the V_N-B, Cy-CN catalyst under visible light irradiation, the concentration of NH_4^+ increased steadily with irradiation time, and the rate reaching 115.53 μmol/g·h, being about 25.5 times that of the pristine g-CN (4.53 μmol/g·h). The N_2-fixation activity was very stable after 10 h of reaction.

S-doped g-CN, which was synthesized by calcination of thiourea in a small open crucible, was placed in a large crucible with a lid at different temperatures (450°C, 500°C, and 550°C) for 4 h [3]. The resultant catalysts obtained from the large crucible were S-doped, but simultaneously contained C-vacancies V_C, and coded as V_C-S-CN-X, where X corresponds to the calcination temperature. The catalyst obtained from the small crucible was S-doped bulk CN (b-S-CN). Their N_2-fixation performances were measured by dispersing 20 mg catalyst powder in an ethanol aqueous solution (4 vol% methanol) under the irradiation of a simulated solar light from a 500 W Xe lamp and the continuous bubbling of air or Ar. NH_3-generation performance was markedly enhanced in the presence of composite V_C-S-CN-*X*s, higher than the bulk *b*-*S*-CN. The NH_3-generation rate reached the highest 5.99 mmol/g·h for the composite Vc-S-CN-550, about 2.8 times that of *b*-S-CN (2.13 mmol/g·h).

An I-doped *g*-CN with C-vacancies (V_C-I-CN) was synthesized by an in situ decomposition-thermal polymerization of a mixture of DCDA with KIO_3 at 550°C for 4 h [4]. The resultant catalysts were coded as V_C-I_X-CN, where *X* is the weight percentage of KIO_3 used (2%, 4%, 6%, 8%, 10%, and 15%). The N_2-fixation performances were studied by dispersing 20 mg catalyst powder in 30 mL of water containing 5 vol% methanol. The V_C-I_{10}-CN exhibited excellent performance and sufficient stability for the photocatalytic N_2 fixation under simulated sunlight irradiation within 3 h (200.8 mg/L·g), being 2.8 times as high as that of the pristine *g*-CN.

A P-doped *g*-CN containing surface N vacancies (V_N-P-CN), which was prepared by simple thermal condensation of a mixture of DCDA with hydroxyethylidene diphosphonic acid (HEDP, $C_2H_8O_7P_2$) in H_2 [5]. It exhibited N_2 fixation, i.e., NH_3-formation from N_2 and water under visible light irradiation. About 5 μmol NH_3 was formed by dispersing the catalyst V_N-P-CN (200 mg) in 100 mL water under N_2 bubbling and the light ($\lambda > 420$ nm) irradiation for 24 h at 30°C.

A novel *g*-CN-based photocatalyst with a donor–acceptor structure was synthesized by copolymerization of DCDA with different amounts of 5-amino-1,10-phenanthroline (AP) at 550°C for 3 h, followed by calcination at 500°C for 2 h in N_2 [6]. In the resultant composite (AP_X-CN), where *X* corresponds to AP-content (0.0–2.0 wt%), AP is covalently linked with CN, as shown in Figure 6.3a. Onto the $AP_{0.5}$-CN, Fe^{3+} was further loaded by mixing the catalyst powder in $FeCl_3$ solution to produce Fe^{3+}-$AP_{0.5}$-CN. The photoluminescence (PL) spectrum of $AP_{0.5}$-CN is compared with the pristine CN and after Fe^{3+}-loading in Figure 6.3b, revealing marked decreases in the intensity of PL peak at 560 nm with the addition of AP and further Fe^{3+}-loading. In Figure 6.3c, the NH_4^+-production rate is shown as a function of *X*. With increasing *X* from 0.0 (i.e., the pristine CN) up to 0.5, NH_3-yield increases from 97.5 to 517.3 μmol/g·h.

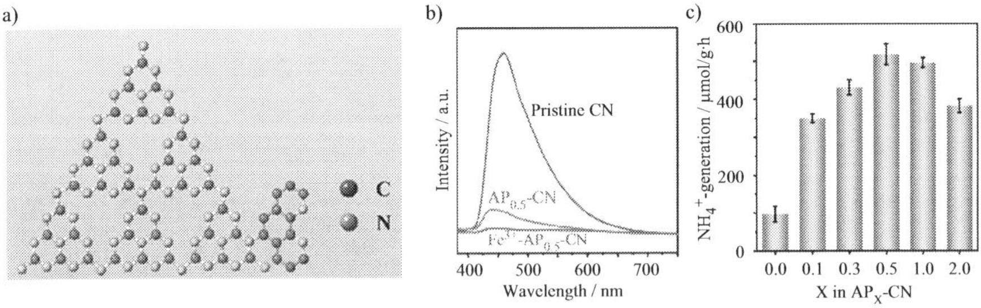

FIGURE 6.3 A *g*-CN-based photocatalyst with a donor-acceptor structure (AP_X-CN): (a) its molecular structure, (b) photoluminescence spectra of $AP_{0.5}$-CN before and after Fe^{3+}-loading in comparison with the pristine CN, and (c) NH_4^+-production rate over AP_X-CN photocatalysts [6].

6.2 $g\text{-}C_3N_4$-BASED COMPOSITES

The self-composite AOP-CN, composed of ternary homojunctions between the amino-rich CN (A-CN), oxidized CN (O-CN), and phosphorylated CN (P-CN), delivered high performance for N_2 fixation [7]. The composite AOP-CN was converted to a hydrogel with the assistance of pectin. Photocatalytic N_2 fixation over the AOP-CN composite hydrogel in comparison with other related CNs is shown as NH_4^+-production kinetics in water with or without a sacrificial agent (methanol) in Figure 6.4a and b, respectively, revealing a marked enhancement of NH_4^+ production on AOP-CN composite, particularly after its hydrogelation. Ammonia production reached 404 μmol in 60 min of visible light irradiation without a sacrificial agent, while increasing to 712 μmol upon adding 10 wt% methanol as a sacrificial agent. N_2-fixation performance of the AOP-CN hydrogel exhibited a slight decrease in ammonia production from 404 to 371 μmol after 5 cycles, as shown in Figure 6.4c.

B-doped CN (B-CN) was loaded by Ag nanoparticles with different Ag contents (1–4 wt%) [8], coded as Ag_X/B-CN, where X corresponds to the content of Ag nanoparticles. UV-vis diffused reflectance spectra are compared on the resultant catalysts in Figure 6.5a. By B-doping (0.5%), the spectrum had a slight red-shift from

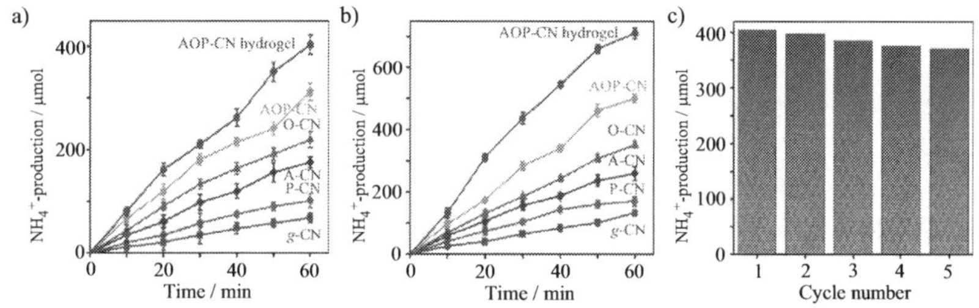

FIGURE 6.4 Photocatalytic N_2 fixation over a self-composite AOP-CN and its hydrogel in comparison with other related CNs: (a) without sacrificial agent, (b) with methanol (sacrificial agent), and (c) reusability [7].

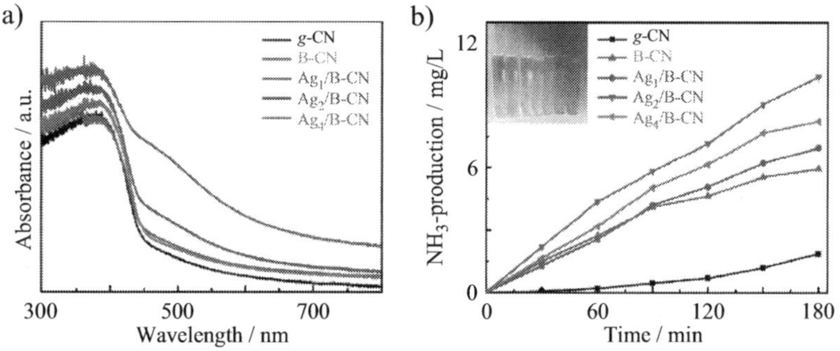

FIGURE 6.5 Ag-loaded B-doped g-CN (Ag_X/B-CNs) in comparison with the pristine g-CN and B-doped CN (B-CN): (a) UV-vis absorbance spectra and (b) NH_3-production kinetics under simulated sunlight irradiation, and digital photographs (inset) of the solution during the photocatalytic process over Ag_2/B-CN [8].

the pristine *g*-CN and the absorption in the visible light region increased with the loading of Ag nanoparticles. The bandgap was calculated to be 2.82, 2.74, 2.73, 2.71, and 2.58 eV for the pristine *g*-CN, B-CN, Ag_1/B-CN, Ag_2/B-CN, and Ag_4/B-CN, respectively. NH_3-production kinetics are shown in Figure 6.5b, revealing that the B-doping is favorable for enhancing N_2 fixation and loading of Ag nanoparticles is also up to 2 wt% loading. Excess loading of Ag nanoparticles on the surface of CN nanosheets may act as the recombination center of photogenerated electron-hole pairs and influence the light transportation to the solution.

Ru quantum dots (RuQDs) with about 3 nm size could significantly enhance the photocatalytic N_2-fixation performance of *g*-CN by loading RuQDs ($RuQD_X$/CN, where X corresponds to Ru-loading) (1, 3, and 5 wt%) [9]. $RuQD_3$/CN exhibited the highest NH_3-production rate of 20.55 μmol/g after 2 h of visible light irradiation in pure water.

N_2-fixation mechanism was discussed on a layer-by-layer (2D/2D) composite of Ti_3C_2 (MXenes) nanosheets and *g*-CN with N-vacancies (V_N-CN) [10]. Into the dispersed CN nanosheets solution (1 mg/mL), a certain volume of the Ti_3C_2 nanosheets solution (1 mg/mL) was added with sonication and stirring for 30 min under N_2 flow to ensure a homogeneous dispersion of the Ti_3C_2 (TC) and CN nanosheets. The mass percent of TC to CN was controlled from 0.5% to 10%. After centrifugation (0–10,000 rpm) and following vacuum drying, the obtained product was calcined in N_2 at a temperature range of 150–300°C for 2 h. The resultant composites were coded as TC_X/V_N-CN, where X corresponds to the mass percentage of TC to CN. NH_3-yield (N_2-fixation performance) depends on the TC mass percentage, calcination temperature, and centrifugation condition, as shown in Figure 6.6a–c, respectively. The optimum conditions were determined to be 2%, 200°C and 10,000 rpm centrifugation, respectively. The composite prepared by this optimized condition was coded as TC_2/V_N-CN, which exhibited the NH_3-formation rate of 5.792 mg/g·h. NH_3-yield increases rapidly with reaction time under N_2 flow for TC_2/V_N-CN. The strong effect of centrifugation, i.e., the composite obtained by 10,000 rpm centrifugation, gave a high NH_3-yield, even 5 times higher than the composite obtained without centrifugation, indicating the edge Ti atoms of the TC nanosheets are the active sites for the N_2-fixation reaction.

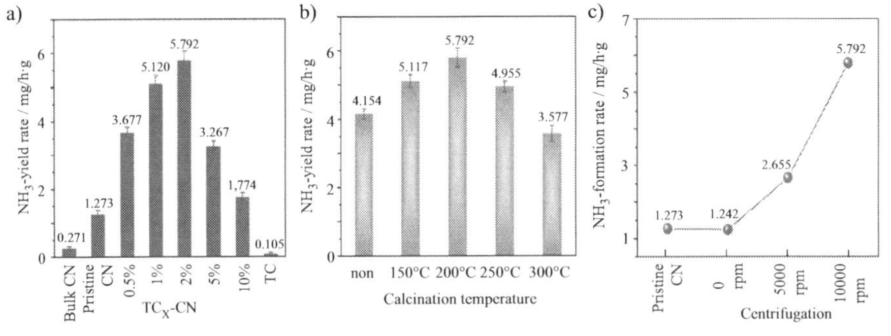

FIGURE 6.6 N_2-fixation performances of Ti_3C_2-loaded *g*-CN with N-vacancies (TC_X/V_N-CN) layer-by-layer composite: (a) the effect of mass% of TC to CN on NH_3-yield, (b) the effect of calcination temperature on NH_3-yield, and (c) effect of centrifugation [10].

The composites of g-CN nanosheets (*ns*-CN) with black phosphorous nanosheets (*ns*-BP) were fabricated by mixing the respective aqueous dispersions (0.3 mg/mL concentration) under sonication in Ar [11]. The mixture was centrifuged, dried under vacuum, and then calcinated at 300°C for 2 h in Ar to synthesize *ns*-BP loaded *ns*-CN composite (coded as BP_X/CN, where X corresponds to the mass ratio of *ns*-BP/*ns*-CN), which was composed of C–P bonds. The *ns*-CN and *ns*-BP were prepared through liquid exfoliation of the DCDA-derived b-CN and bulk black phosphorous, respectively. The N_2-fixation performances were measured by dispersing composite catalyst powder in 50 mL of water with 5 vol% methanol under visible light irradiation and N_2 bubbling. The composite $BP_{0.05}$/CN exhibited the highest performance, NH_3-production rate reaching 347.5 µmol/L·h, while the 0.05 wt% Pt-loaded CN ($Pt_{0.05}$/CN) delivered almost comparable NH_3-production rate (337.2 µmol/L·h) to the resultant composite $BP_{0.05}$/N.

The composites of g-CN with transition metal oxides (Fe_2O_3, VO_2, In_2O_3, ZrO_2, and CrO_3) were prepared [12] by heating the mixtures of urea and melamine with different amounts of the metal chloride salts at 95°C for a few minutes to form deep eutectic solvents, and by following calcination at 550°C for 4 h. The N_2-fixation performances of these composite photocatalysts were studied by dispersing the catalyst powders in a water/methanol (9/1 by volume) mixture under UV-Vis light irradiation from a 300 W Xe lamp. N_2-fixation (NH_3-generation) behaviors are shown on the composite with different amounts of Fe_2O_3, which were coded as $(Fe_2O_3)_X$/CN, where X corresponds to 30–50 wt% of Fe_2O_3. The NH_3-generation kinetics are shown on the composites loaded with Fe_2O_3 in different amounts, and those on the composites with other transition metals are shown in Figure 6.7a and b, respectively. On the composites containing Fe_2O_3, the composite $(Fe_2O_3)_{40}$/CN exhibited the highest NH_3-generation even in comparison with the composites with other transition metals. Its NH_3-generation rate reached 4.38 mmol/L·h, as shown in Figure 6.7c.

A composite TiO_2@C/g-CN was prepared via one-step calcination of a mixture of Ti_3C_2 and melamine (melamine/Ti_3C_2mol ratio of 0/1, 10/1, 20/1, and 50/1) at 550°C for 2 h [13]. During calcination at 550°C, $Ti_3C_2T_x$ MXene was converted to TiO_2, which was anchored on a carbon sheet (TiO_2@C) and melamine to g-CN, resulting in a composite TiO_2@C/g-CN. The NH_3-generation performance was studied by

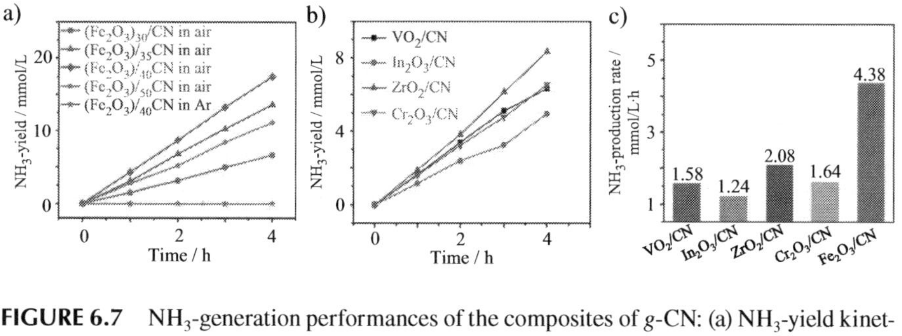

FIGURE 6.7 NH_3-generation performances of the composites of g-CN: (a) NH_3-yield kinetics for the composites $(Fe_2O_3)_X$/CN, (b) those for the composite with other transition metal oxides, and (c) NH_3-production rate for the composites of different metal oxides [12].

dispersing 50 mg catalyst powder in 100 mL of distilled water containing 20 vol% methanol under pure N_2 bubbling and irradiation of visible light from a 300 W Xe lamp. The composite obtained using melamine/Ti_3C_2 of 10/1 delivered the highest N_2-fixation (NH_3-production) performances, giving a rate of 250.6 μmol/g·h, higher than those of the pristine *g*-CN and TiO@C, and the mechanical mixture of *g*-CN with TiO@C. NH_3-production was not detected in the absence of a nitrogen atmosphere and without visible light irradiation.

A composite of melamine-derived *g*-CN nanosheet obtained by exfoliation with different amounts of bismuth molybdate (Bi_2MoO_6, BMO) was synthesized via a one-step reflux process of a mixed aqueous solution of CN, $Bi(NO_3)_3 \cdot 5H_2O$, and $Na_2MoO_4 \cdot 2H_2O$ at 96°C for 2 h [14]. On the resultant BMO_X/CN composites (*X* corresponds to the BMO-content), N_2-fixation performances were investigated by dispersing 40 mg of the composite in water containing absolute ethanol as a hole scavenger (0.789 g/L) under the light from a 500 W Xe lamp with a 420 nm cutoff filter (power density of 100 mW/cm^2). UV-vis absorbance spectra of the composites and NH_3-yield measured by NH_4^+ concentrations are compared on the composites together with the pristine *g*-CN and BMO in Figure 6.8a and b, respectively. All are UV-vis active and gave a high NH_3-yield. The exfoliation of *g*-CN is effective for the enhancement of NH_3-yield. In addition, the compositing of *g*-CN with photocatalyst BMO promotes the NH_4^+ formation; the BMO_{30}/CN composite giving the highest NH_4^+ production rate of 3,271 μmol/L·g.

A photocatalyst, 3,4-dihydroxybenzaldehyde(DBD)-functionalized Ga_2O_3/*g*-CN having heterojunctions, was synthesized and coded as $(Ga_2O_3\text{-DBD})_X$/CN (*X* corresponds to the weight percentage of Ga_2O_3) by direct thermal polymerization of the mixtures of urea with Ga_2O_3-DBD nanoparticles by calcination at 550°C [15], which delivered a high N_2-fixation performance. With the increase in the content of Ga_2O_3-DBD nanoparticles, the composite showed an improvement of the light absorption in both the UV and visible regions compared with the pristine *g*-CN, as shown in Figure 6.9a. The photocatalytic N_2-fixation activities of $(Ga_2O_3\text{-DBD})_X$/CN composite were compared with the pristine *g*-CN in Figure 6.9b. The composite $(Ga_2O_3\text{-DBD})_{2.4}$/CN presented the highest photocatalytic activity with a N_2-fixation rate of 112.5 μmol/L·h, 3.37 times higher than that of the pristine *g*-CN (33.4 μmol/L·h),

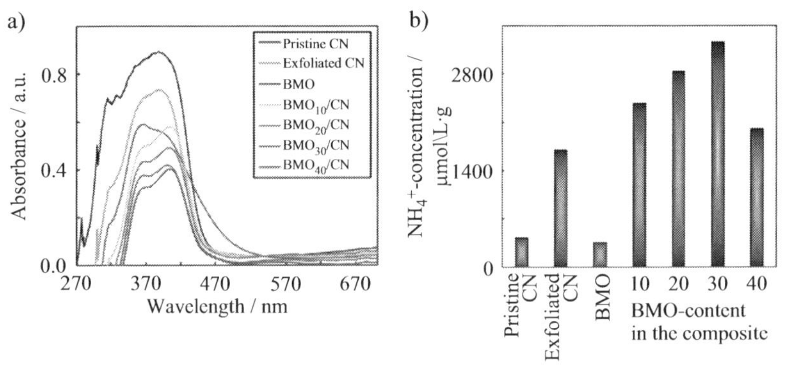

FIGURE 6.8 The composites of *g*-CN with Bi_2MoO_6 (BMO), BMO/CNs: (a) UV-vis absorbance spectra and (b) NH_3-yield (evaluated by NH_4^+-concentration) [14].

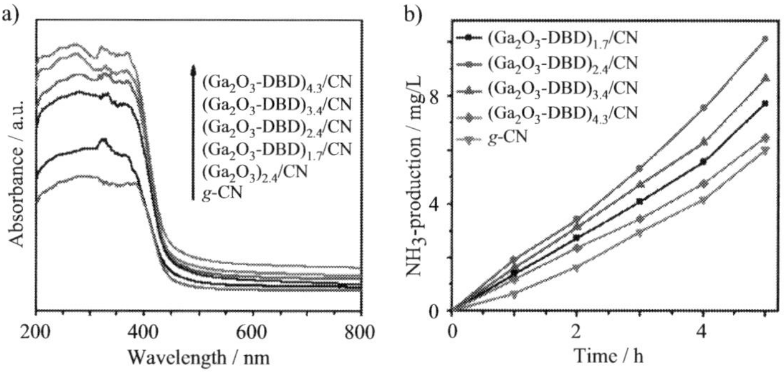

FIGURE 6.9 3,4-dihydroxybenzaldehyde (DBD)-functionalized Ga_2O_3/g-CN, (Ga_2O_3-DBD)$_X$/CN having heterojunctions: (a) UV-vis absorbance spectra and (b) N_2-fixation performance (NH_3-production kinetics) [15].

A composite of rare earth-doped g-CN loaded with a UV-activated Ce-UiO-66 MOF particle (RE-CN/actMOF) exhibited efficient N_2-fixation performances [16]. Rare earth-doped g-CN (RE-CN) was synthesized by heating the mixture of 5 g melamine with 5 wt% $RE(NO_3)_3$ (RE: La, Nd, Sm, or Yb) at 550°C for 3 h. Among the composites using different RE-doped CNs, Sm-CN/actMOF composite exhibited the highest visible-light absorption and efficient utilization of photocarriers, reaching an apparent quantum efficiency (AQE) of 1.58% under 495 nm light irradiation. To stabilize the active sites in UV-activated MOF particles, a graphene sheet was embedded in the MOF particles. Graphene-embedding delivered high stability of UiO-66 MOF and high N_2-fixation performances (NH_3-production of 3.06 mmol/L in 24 h over 20 mg of catalyst) [17]. The photocatalyst $Co(OH)_2$ nanosheets were loaded onto a MOF (NH_2-MIL-125), which could reduce nitrate ions to ammonia (rate of 2.455 mmol/g·h), accompanying a suppression of the competitive H_2 evolution [18].

The Nb-doped g-CN composed of half-curled worm-like nanosheets exhibited a high sensitivity for NH_3 detection [19]. The g-CN was synthesized from a mixture of urea (15 g) and melamine (3 g) calcined at 550°C for 3 h in a semi-closed crucible, onto which Nb nanoparticles were loaded by mixing with different amounts of niobium chloride, $NbCl_5$ (100, 200, and 300 mg), and following heating at 550°C for 3 h. Nb-CN prepared using 200 mg $NbCl_5$ exhibited a response time of about 6 sec and a 90% recovery time of about 29 sec toward 50 ppm ammonia gas exposure with 76% sensitivity. The composite of g-CN with a perovskite $GdInO_3$ also exhibited a high sensitivity for NH_3 detection [20]. The composite was prepared by calcination of a mixture of melamine (3 g) with $GdInO_3$ (1.5 g) at 550°C for 5 h in a ceramic crucible. It exhibited an impressive response of 72% toward 10-ppm ammonia, coupled with a response time of 63 sec and a recovery time of 33 sec at room temperature.

The CN framework was functionalized by a benzene ring and polychlorination, which was achieved through thermal copolymerization of 20 g urea with X mg 2,2′,5,5′-tetrachlorobenzidine (TB) (X = 15, 30, 45, and 60) at 550°C for 3 h (TB_X-CN) [21]. The synthesis process is illustrated in Figure 6.10a. The resultant TB_X-CN composites were applied to the study of N_2-fixation performance by dispersing 20 mg TB_X-CN in 100 mL deionized water containing 5 vol% triethanolamine as a sacrificial agent and

H_2PtCl_6 aqueous solution (1 wt% Pt) as a co-catalyst under simulated solar light from a 300 W Xe lamp. N_2-fixation performance of TB_X-CN composite catalysts is shown in Figure 6.10b, revealing that compositing CN with TB enhanced N-fixation ability markedly, the TB_{45}-CN exhibiting the highest NH_3-production. The resultant composites delivered high performance for the photodegradation of tetracycline.

6.3 C_3N_5-BASED MATERIALS

The C_3N_5 containing square-holes (macropores) (*sh*-C_3N_5) was synthesized using KBr-templating at 550°C, and its N_2-fixation performances were studied [22], of which the SEM image and molecular structure are shown as the insets in Figure 6.11a and b, respectively. N_2-fixation performance of *sh*-C_3N_5 using 50 mg catalysts was compared with the pristine C_3N_5 synthesized without KBr and their composites with V-doped carbon dots (V-CD). V-CD was synthesized from chitosan oligosaccharide with 1,10-phenanthroline ($C_{12}H_8N_2$) and V(IV)-oxy acetylacetonate ($C_{10}H_{14}O_5V$) via the solvothermal procedure at 160°C for 12 h. In Figure 6.11a, UV-vis spectra of *sh*-C_3N_5 and its composite with V-CD (V-CD/C_3N_5 and V-CD/*sh*-C_3N_5) are shown, suggesting that both are UV-vis active. In Figure 6.11b, the NH_3-yielding was shown on *sh*-C_3N_5 and the composite V-CD/*sh*-C_3N_5 in comparison with the pristine C_3N_5 and V-CD/C_3N_5 composite, revealing marked enhancements in NH_3-yield by introducing square-holes (square-shaped macropores) in C_3N_5 (preparation of the *sh*-C_3N_5) and compositing with V-CD.

6.4 CONCLUDING REMARKS

The N_2-fixation performances reported are summarized by focusing on NH_3-generation rate together with synthesis conditions in Table 6.1. Since the NH_3-generation rate reported in different units in literature referred, the values recalculated on the basis of measurement conditions of N_2-fixation performances are listed in Table 6.1.

The *g*-CNs containing N-vacancy (V_N) and C-vacancy (V_C) have been studied, but a marked enhancement in NH_3-generation rate was not obtained. On Table 6.1, the introduction of O-vacancies in *g*-CN structure looks to enhance NH_3-generation reaction [3,4]. Here, it is difficult to conclude what the key is for the enhancement of NH_3-generation rate.

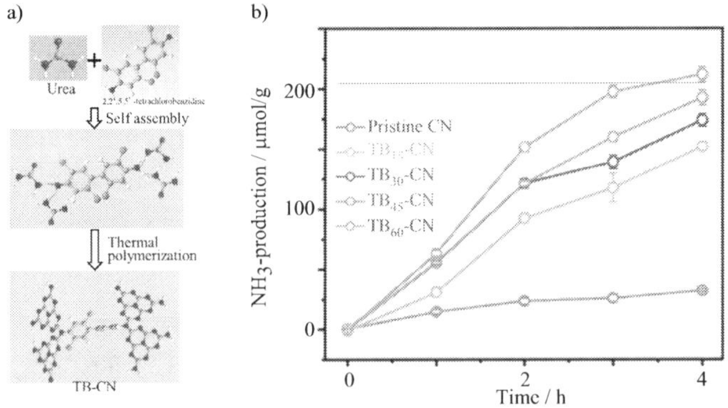

FIGURE 6.10 N_2-fixation performances of TB_X-CN composite catalysts [21].

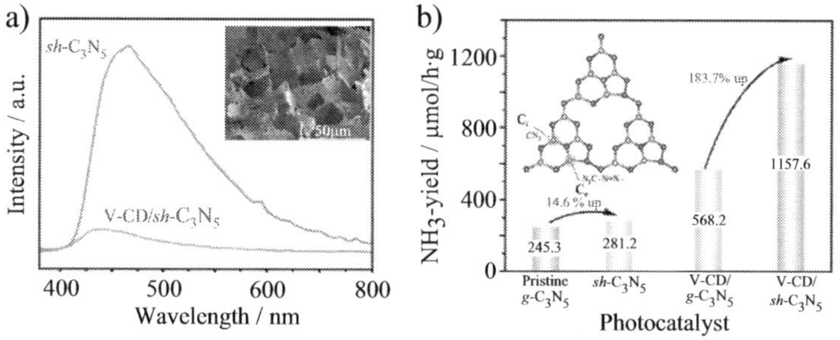

FIGURE 6.11 The photocatalytic performances of the composites based on g-C_3N_5: (a) UV-vis spectra of sh-C_3N_5 and the composite V-CD/sh-C_3N_5 with a SEM image (inset), and (b) NH_3-yield by different photocatalysts with the structure of C_3N_5 (inset) [22].

TABLE 6.1
N_2-Fixation (NH_3-Generation) Performances of *g*-CN- and C_3N_5-Based Photocatalysts

		N_2 Fixation (NH_3 Generation)		
Photo-Catalyst	**Synthesis and Modification of *g*-CN**	**Co-Catalyst, etc.**	**Rate (mmol/ g·h)**	**References**
	***g*-C_3N_4 and Modified *g*-C_3N_4**			
V_N-CN-4	Melamine-derived *g*-CN, further 620°C for 4 h, formation of N-vacancies (V_N)	Potentiostatic measurement, 0.1 M HCl solution	0.476[a]	1
V_N-BCy-CN	B and cyano-group (Cy) co-doping, formation of N-vacancies (V_N)	50 mg catalyst in 50 mL water with stirring in N_2	0.116	2
Vc-S-CN-550	S-doping at 550°C, simultaneous formation of C-vacancies (V_C)	20 mg catalyst in water with 4 vol% methanol	5.99	3
b CN	Pristine bulk *g*-CN		2.13	
V_C-I_{10}-CN	I-doping (DCDA + 10 wt% KIO_3, 550°C), simultaneous V_C formation	20 mg catalyst in 30 mL water with 5 vol% methanol	5.13	4
$AP_{0.5}$-CN	DCDA + 0.5 wt% 5-amino-1,10-phenanthroline (AP, 500°C)	10 mg catalysts in 100 mL water with 10 vol% methanol	0.517	6

(*Continued*)

TABLE 6.1 (*Continued*)
N_2-Fixation (NH_3-Generation) Performances of *g*-CN- and C_3N_5-Based Photocatalysts

		N₂ Fixation (NH₃ Generation)		
Photo-Catalyst	**Synthesis and Modification of *g*-CN**	**Co-Catalyst, etc.**	**Rate (mmol/ g·h)**	**References**
	***g*-C_3N_4-Based Composites**			
AOP-CN hydrogel	Self-composite of amino-rich CN (A-CN), oxidized CN (O-CN), and phosphorylated CN (P-CN)	Pure water	1.347[a]	7
		Water with 10 wt% methanol	2.373[a]	
Ag_2/B-CN	2 wt% Ag loading onto B-doped CN (B-CN)	30 mg catalyst in 40 mL of water with 5 vol% methanol	0.305[a]	8
TC_2/V_N-CN	2 wt% Ti_3C_2 (TC) nanosheets-loading onto V_N-CN at 200°C, 2D/2D heterojunction	50 mg catalyst in 40 mL of water with 10 mL of methanol	0.341[a]	10
$BP_{0.05}$/CN	0.05 mass% black phosphorus (BP)-loading onto *ns*-CN	30 mg catalyst in water with 5 vol% methanol	0.579[a]	11
$Pt_{0.05}$/CN	0.05 mass% Pt-loading onto *ns*-CN		0.562[a]	
$(Fe_2O_3)_{40}$/CN	Calcination of deep eutectic solvent, $FeCl_3$/urea/melamine, 550°C	Water with 10 wt% methanol	4.38	12
TiO_2@C/CN	$Ti_3C_2T_x$ converted to TiO_2@C on *g*-CN nanosheet	50 mg catalyst in water with 20 vol% methanol N_2 bubbling	0.2506[a]	13
Ga_2O_3-$DBD_{2.4}$/ *g*-CN	2.4% dihydroxybenzaldehyde(DBD)-functionalized Ga_2O_3-loading, urea	20 mg catalyst in water with 0.04 mM methanol	0.281[a]	15
	C_3N_5-Based Materials			
V-CD/*g*-C_3N_5	V-doped carbon dots-loading onto *g*-CN	Water + 20 vol% methanol at 35°C pure N_2 bubbling	0.568	22
V-CD/*sh*-C_3N_5	onto square-hole *g*-C_3N_5 (*sh*-CN) synthesized by KBr-templating		1.158	

[a] Recalculated to the common unit (mmol/g·h) by taking into account the experimental conditions.

REFERENCES

1. Lv C, Qian Y, Yan C, et al., Defect engineering metal-free polymeric carbon nitride electrocatalyst for effective nitrogen fixation under ambient conditions. *Angew Chem Int Ed* 2018; 57: 10246–50.
2. Sun D, Li L, Yu Y, et al., B and cyano groups co-doped g-C_3N_4 with multiple defects for photocatalytic nitrogen fixation in ultrapure water without hole scavengers. *J Colloid Interface Sci* 2021; 600: 639–48.
3. Cao S, Fan B, Feng Y, et al, Sulfur-doped g-C_3N_4 nanosheets with carbon vacancies: General synthesis and improved activity for simulated solar-light photocatalytic nitrogen fixation. *Chem Eng J* 2018; 353: 147–56.
4. Hu X, Zhang W, Yong Y, et al., One-step synthesis of iodine-doped g-C_3N_4 with enhanced photocatalytic nitrogen fixation performance. *Appl Surf Sci* 2020; 510: 145413.
5. Shiraishi Y, Shiota S, Kofuji Y, et al., Nitrogen fixation with water on carbon-nitride-based metal-free photocatalysts with 0.1% solar-to-ammonia energy conversion efficiency. *ACS Appl Energy Mater* 2018; 1: 4169–77.
6. Cui D, Yang X, Liu Y, et al., Strategy of constructing D-A structure and accurate active sites over graphitic carbon nitride nanowires for high efficient photocatalytic nitrogen fixation. *J Colloid Interfaces Sci* 2025; 678(B): 955–69.
7. Balakrishnan A, Chinthala M, Kumar A, et al., Construction of dual Z-scheme ternary carbon nitride homojunction pectin microspheres as a multifunctional photocatalyst for tetracycline degradation, H_2O_2 production, and N_2 fixation. *Chem Eng J* 2024; 496: 153899.
8. Yao X, Zhang W, Huang J, et al., Enhanced photocatalytic nitrogen fixation of Ag/B-doped g-C_3N_4 nanosheets by one-step in-situ decomposition-thermal polymerization method. *Appl Catal A* 2020; 601: 117647.
9. Wang S, Guo D, Zong M, et al., Unravelling the strong metal-support interaction between Ru quantum dots and g-C_3N_4 for visible-light photocatalytic nitrogen fixation. *Appl Catal A* 2021; 617: 118112.
10. Sun C, Chen Z, Cui J, et al., Site-exposed Ti_3C_2 MXene anchored in N-defect g-C_3N_4 heterostructure nanosheets for efficient photocatalytic N_2 fixation. *Catal Sci Technol* 2021; 11: 1027–38.
11. Qiu P, Xu C, Zhou N, et al., Metal-free black phosphorus nanosheets-decorated graphitic carbon nitride nanosheets with C–P bonds for excellent photocatalytic nitrogen fixation. *Appl Catal B* 2018; 221: 27–35.
12. Mou H, Wang J, Zhang D, et al., A one-step deep eutectic solvent assisted synthesis of carbon nitride/metal oxide composites for photocatalytic nitrogen fixation. *J Mater Chem A* 2019; 7: 5719–25.
13. Liu Q, Ai L, Jiang J, MXene-derived TiO_2@C/g-C_3N_4 heterojunctions for highly efficient nitrogen photofixation. *J Mater Chem A* 2018; 6: 4102–10.
14. Vesali-Kermani E, Habibi-Yangjeh A, Diarmand-Khalilabad H, et al., Nitrogen photofixation ability of g-C_3N_4 nanosheets/Bi_2MoO_6 heterojunction photocatalyst under visible-light illumination. *J Colloid Interfaces Sci* 2020; 563: 81–91.
15. Cao S, Zhou N, Gao F, et al., All-solid-state Z-scheme 3,4-dihydroxybenzaldehyde-functionalized Ga_2O_3/graphitic carbon nitride photocatalyst with aromatic rings as electron mediators for visible-light photocatalytic nitrogen fixation. *Appl Catal B: Environ* 2017; 218: 600–10.
16. Zhu S, Liu S, Zhang L, et al., Rare earth-doped carbon nitride composite Ce-based metal-organic framework for visible light-driven nitrogen fixation catalysis. *J Rare Earths* 2025; 43: 1607–15.
17. Liu S, Teng Z, Liu H, et al., A Ce-UiO-66 metal–organic framework-based graphene-embedded photocatalyst with controllable activation for solar ammonia fertilizer production. *Angew Chem Int Ed* 2022; 61: e202207026.

18. Xi Y, Xiang Y, Bao T, et al., Nanoarchitectonics of S-scheme heterojunction photocatalysts: A nanohouse design improves photocatalytic nitrate reduction to ammonia performance. *Angew Chem Int Ed* 2024; 63: e202409163.
19. Sethuraman S, Marimuthu A, Kattamuthu R, et al., Highly surface active niobium doped g-C_3N_4/g-C_3N_4 heterojunction interface towards superior photocatalytic and selective ammonia response. *Appl Surf Sci* 2021; 561: 150077.
20. Amith SL, Radhakrishnan K, Gurunathan K, Utilizing e-waste: development of a room temperature ammonia sensing device featuring highly crystalline $GdInO_3$ nanoballs on surface-active g-C_3N_4 nanosheets. *J Alloys Compd* 2024; 1005: 175930.
21. Chen Y, Shi J, Xu Y, et al., Polychloride and benzene ring synergistically boosting photocatalytic degradation and nitrogen fixation performances of carbon nitride. *Sep Purif Technol* 2025; 354: 129231.
22. Cai W, Li K, Zhu Z, et al., Construction of microreactor-like square-hole carbon nitride boosting photocatalytic nitrogen fixation activity. *Ind Eng Chem Res* 2024; 63(29): 12852–62.

7 Pollutant Degradation and Removal by *g*-C_3N_4

Degradation (or mineralization) and removal of environmental pollutants are important and urgent challenges in maintaining a clean and safe environment. *g*-C_3N_4-based materials are attracting attention mainly because of their visible-light (solar light) active photocatalytic activities, which can be achieved through low-cost and environment-friendly synthesis processes. Consequently, these materials are expected to make significant contributions to the environmental remediation. A comprehensive review on *g*-CN as an emerging photocatalyst for sustainable environmental applications was studied [1].

In this chapter, the *g*-CN-related photocatalysts are classified, as in previous chapters, into two: *g*-CN and modified *g*-CNs and *g*-CN-based composites. Since the pollutants used with *g*-CN-related materials are so varied, the results reported were presented by classifying them according to the pollutants, such as organic dyes, antibiotics, emerging chemicals, and bacteria.

7.1 *g*-C_3N_4 AND MODIFIED *g*-C_3N_4

7.1.1 Organic Dyes

2D layered *g*-CN synthesized at 550°C for 3 h exhibited an enhanced photocatalytic activity in the removal of various organic dyes [2]. The photodegradation kinetic curves of various dyes, such as Rhodamine B (RhB), Crystal Violet (CV), Methylene Blue (MB), Methyl Orange (MO), Naphthol Orange (NO), and a phenol derivative, *p*-nitrophenol (*p*-NP), by suspending *g*-CN particles under sunlight irradiation are shown in Figure 7.1a. These kinetics could be approximated by zero-order kinetic model and the rate constant k is calculated to be 1×10^{-2}, 1.1×10^{-2}, 9.6×10^{-3}, 6.7×10^{-3}, 1.7×10^{-2}, and 4.6×10^{-2}/min for MB, RhB, CV, NO, MO, and *p*-NP respectively. In Figure 7.1b, cyclic applicability of RhB-degradation is shown in three cycles, revealing the catalyst's cycle stability. Rapid degradation of Rh-6G dye over urea-derived *g*-CN was reported, 99.6% degradation within 300 min under UV irradiation [3].

The photocatalytic activity for the RhB degradation under visible light was evaluated on a mesoporous *g*-CN (*mpg*-CN), which was derived from melamine/cyanuric acid complex (MC-derived CN), as well as *g*-CN derived from dicyandiamide (DCDA-derived CN) [4]. The changes in relative concentration of RhB with irradiation time are shown in Figure 7.2a. The RhB-degradation proceeds rapidly with increasing time in the presence of the MC-derived CN, much faster than the DCDA-derived CN and a commercial photocatalyst TiO_2 (P25). The

DOI: 10.1201/9781003589051-7

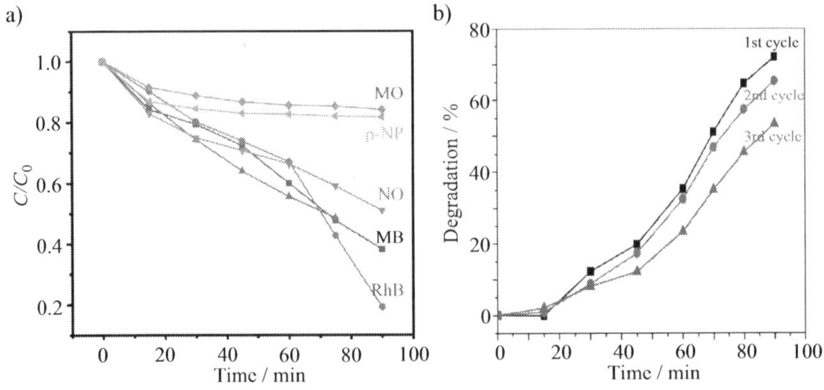

FIGURE 7.1 2D-layered g-CN catalyst: (a) photodegradation of various dyes and (b) cyclability for RhB-degradation [2].

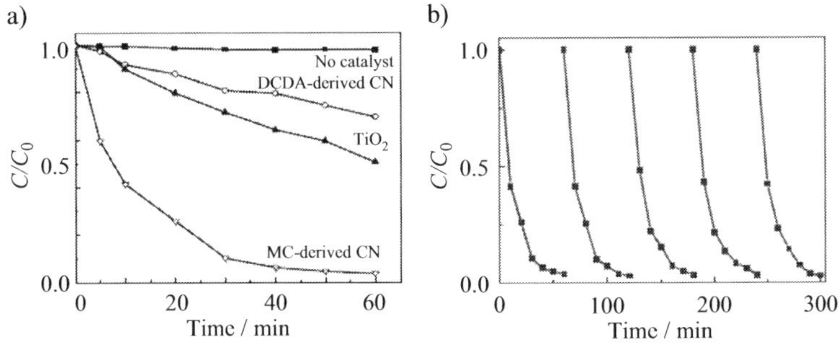

FIGURE 7.2 Photocatalytic degradation of RhB with the existence of the MC-derived CN, DCDA-derived CN, and TiO_2: (a) time change of RhB-concentration with visible light irradiation and (b) cyclability of the MC-derived CN [4].

MC-derived CN catalyst exhibits an excellent cyclability for RhB-degradation, as shown in Figure 7.2b.

The melamine/cyanuric acid complex (MC-complex) forms many different ordered textures, depending on the solvent [5]. The MC-complex was prepared from the mixture of melamine and cyanuric acid (1/1 by mole) using different solvents [6]. The particles obtained from ethanol solution with different calcination times (2–12 h) exhibited a highly ordered pancake-like morphology due to the formation of extended sheets consisting of the hydrogen-bonded supramolecular structure. The resultant MC-complexes were converted to C_3N_4 by calcination at 550°C in N_2. The resultant CNs were photocatalytically active, depending strongly on the calcination time during MC-complex synthesis. In Figure 7.3a, the diffuse reflectance absorption spectra are shown on the MC-derived CNs using different calcination times in comparison with the DCDA-derived CN. In Figure 7.3b, RhB-degradation performances of the MC-derived CNs are shown, together with the DCDA-derived CN. The RhB-degradation activity depends strongly

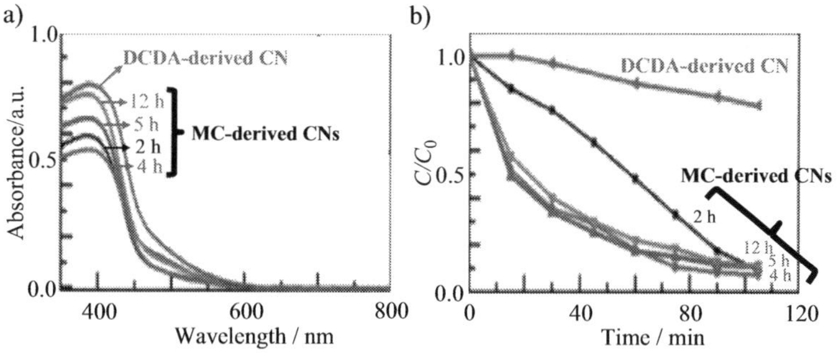

FIGURE 7.3 The CM-derived CNs prepared by using different calcination times in comparison with the DCDA-derived CN: (a) UV-vis absorption spectra and (b) RhB-degradation curves [6].

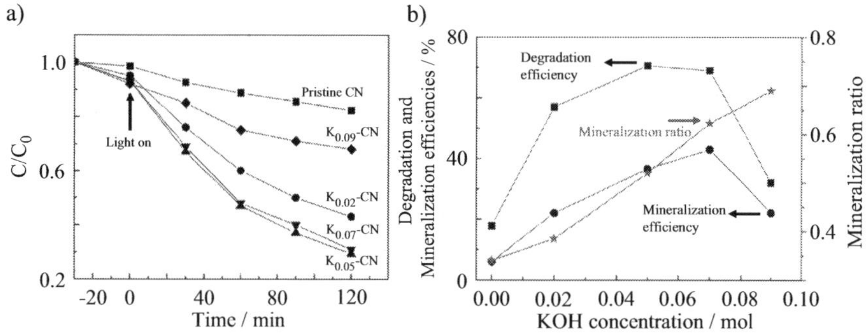

FIGURE 7.4 Photocatalytic degradation performances of RhB by K-doped CNs ($K_{0.02}$-CN–$K_{0.09}$-CN) in comparison with the pristine g-CN: (a) C/C_0 *vs.* irradiation time and (b) changes in degradation and mineralization rates and mineralization ratio with KOH concentration used [8].

on the calcination time, while much higher than that of the DCDA-derived CN. The MC-derived CN could degrade RhB completely after 105 min, while the DCDA-derived CN gave much slower RhB-degradation (less than 20% degradation even after 105 min). DCDA-derived CN by calcination above 500°C worked as a photocatalyst to activate H_2O_2 to degrade RhB in water under visible light [7].

K-doping into g-CN was shown to be effective for the photocatalytic degradation removal of RhB in water [8]. The K-doped CN was prepared from the precipitates, which were obtained from DCDA aqueous solution by adding KOH solution of different concentrations (0.02, 0.05, 0.07, and 0.09 mol/L), by calcination at 520°C. The resultant CNs were coded as K_X-CN, where X represents the KOH concentration used. In Figure 7.4a, the changes in RhB-concentration C/C_0 with irradiation time are shown, revealing a significant enhancement in photocatalytic activity for RhB-degradation removal with K-doping into the CN catalyst. Prior to turning on the light, the RhB adsorption was observed on the K_X-CN catalysts. The $K_{0.05}$-CN catalyst exhibited the highest RhB-degradation performance, while the activity decreased with an increase in KOH concentration. The curves in Figure 7.4a could be well approximated by the first-order relationship, from which the rate constant k for the RhB-degradation was calculated to be 0.0017, 0.0074, 0.011, 0.010, and 0.0037 min^{-1} for the pristine g-CN, $K_{0.02}$-CN, $K_{0.05}$-CN, $K_{0.07}$-CN, and $K_{0.09}$-CN, respectively. The $K_{0.05}$-CN exhibited the highest k, which was 7.4 times higher than that of the pristine g-CN, while the $K_{0.07}$-CN produced nearly identical k-value. The mineralization ability of the resultant catalysts to RhB was evaluated by monitoring the changes of TOC (total organic carbon) in the reaction system. The mineralization efficiency and ratio are shown by plotting against the KOH-concentration used, together with the degradation rate, in Figure 7.4b. The mineralization ratio, which represents the mineralization ability, is equal to the ratio of the mineralization efficiency to the degradation efficiency. Clearly, the mineralization efficiencies of the resultant catalysts show a trend similar to the degradation efficiencies, although $K_{0.07}$-CN delivered the highest mineralization efficiency.

A potassium, sodium-co-doped g-CN (K, Na-CN) was synthesized by mixing melamine with a eutectic KCl/NaCl mixture T_m of 476°C) and calcining at 520°C [9]. The KCl-NaCl (PS)/melamine mass ratio was 15/1, 30/1, and 45/1, and the resultant co-doped CN was coded as $(PS)_X$-CN, where X corresponds to the mixing ratio of (PS)/melamine. Their photocatalytic activity for RhB was evaluated by dispersing 50 mg catalyst powder in 200 mL of a 10 ppm RhB solution under the irradiation of the light from a 250 W high-pressure sodium lamp with a wavelength range of 400–800 nm with air bubbling. The RhB-degradation removal performance was markedly enhanced by K and Na co-doping, as shown in Figure 7.5a with their k-value, revealing that the $(PS)_{30}$-CN shows the highest k, which is 10 times higher than that of the pristine CN. The RhB-degradation and mineralization rates, as well as mineralization ratio, are shown for each catalyst in Figure 7.5b. The mineralization ratio increases evidently with increasing the weight ratio of the eutectic salt (PS) to melamine. The cycle stability in the activity was confirmed on $(PS)_{30}$-CN.

B-doping on g-CN was also shown to be effective for enhancing the photodegradation of RhB and MO in water [10]. The photocatalytic degradation activities for

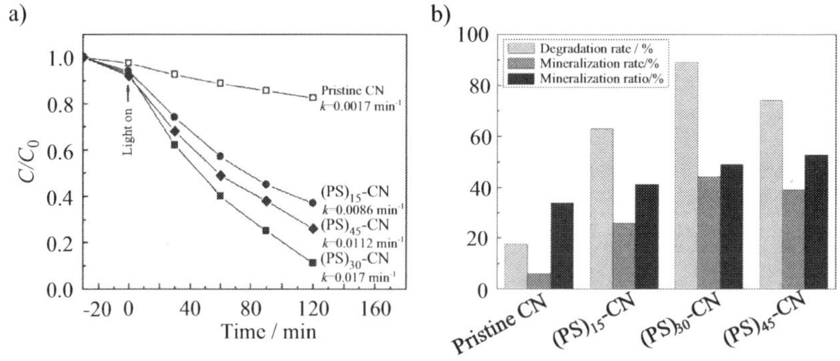

FIGURE 7.5 Photocatalytic degradation performances of RhB dye by K and Na co-doped CNs ($(PS)_{15}$-CN, $(PS)_{30}$-CN, and $(PS)_{45}$-CN) in comparison with the pristine g-CN: (a) C/C_0 vs. irradiation time and (b) degradation and mineralization rates and mineralization ratio for $(PS)_X$-CNs [9].

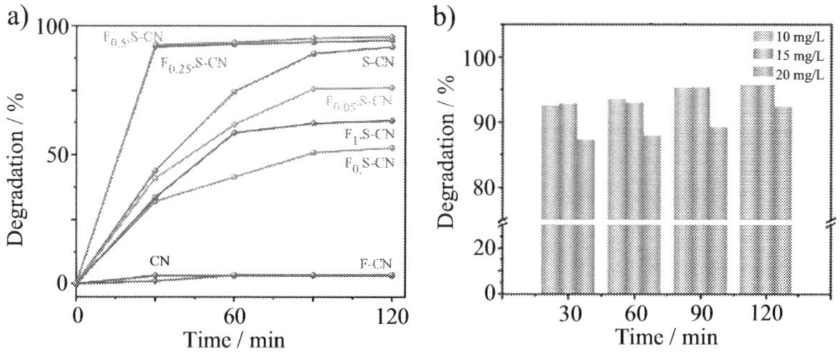

FIGURE 7.6 Degradation removal performances of MO in water by F_X, S-CN (*mpg*-CN) catalysts: (a) degradation curves under light (λ of 405 nm) irradiation and (b) degradation histogram over $F_{0.5}$, S-CN catalyst for different initial concentrations of MO (10, 15, 20 mg/L) [11].

RhB and MO were enhanced on the melamine-derived *g*-CN with additional calcination above 500°C.

Mesoporous *g*-CN (*mpg*-CN) was synthesized from melamine mixing with a 30 wt% colloidal silica (S, hard template) and different amounts of a surfactant F127 (F, soft template) at 550°C for 3 h [11]. The resultant *mpg*-CNs were coded as F_X,S-CNs, where X corresponds to the amount of F127 used ($X = 0$, 0.05, 0.25, 0.5, 1, 1.25, and 1.5 g). For comparison, bulk *g*-CN, silica-templated CN (without F127) and F127-templated CN (without silica) were synthesized from melamine (coded as CN, S-CN, and F-CN, respectively). The $F_{0.5}$, S-CN exhibited the most pore-developed texture, S_{BET} of 104 m^2/g and V_{total} of 0.346 cm^3/g. The photocatalytic activity of the resultant CNs was evaluated by the measurement of MO degradation (10 mg/L concentration) in water under UV irradiation (15 W, 405 nm) for 30 min. The MO-degradation curves under UV irradiation for F_X,S-CNs, in comparison with CN, F-CN, and S-CN, are shown in Figure 7.6a. The modification of *g*-CN using silica templating was effective for the enhancement of MO-degradation. The degradation efficiency of the catalyst S-CN rapidly increased to 91.84% after 90 min of irradiation and then stabilized. However, both $F_{0.25}$, S-CN and $F_{0.5}$, S-CN catalysts prepared using a cooperative organization of hard and soft templates exhibited a much more remarkable degradation efficiency of 91.51% in only 30 min irradiation. As shown in Figure 7.6b, the MO-degradation by the catalyst $F_{0.5}$, S-CN did not decrease when the initial concentration of MO increased to 15 mg/L. When the concentration increased to 20 mg/L, the degradation efficiency dropped slightly but remained about 90%.

An exfoliated *g*-CN (*ex*-CN) exhibited a high degradation rate for hematein (an oxidized form of hematoxylin), which was used in laboratories for cell identification by staining [12]. The degradation rate of hematein over *ex*-CN under solar light irradiation was much faster than that of the bulk *p*-CN. Catalytic performance was improved with an increase in the catalyst concentration. The *ex*-CN exhibited an excellent cyclability over five cycles.

Na- and S-co-doped *g*-CN (Na, S-CN) could detect quercetin by fluorimetry and UV-vis absorption [13]. In comparison to single heteroatom-doped *g*-CN ((Na1)-CN and (S1)-CN), the Na, S-CN exhibited the limit of detection (LOD) for the quercetin by fluorescence and UV-vis absorption, of 23.79 nM and 16.49 μM, respectively, with corresponding linear ranges of 3–80 nM and 0–60 μM, respectively.

A S-doped *g*-CN prepared from thiourea at 550°C in air was treated with diluted H_2SO_4/H_2O_2 aqueous solution (water/acid volume ratio is 7/3) to get acid-activated S-CN [14]. It exhibited a high performance in the degradation of Indigo Carmine (IC), degradation efficiency reaching 96.3% within 40 min in a 25 mg/L IC solution. The IC-degradation reaction rate constant k for the acid-activated S-CN was calculated to be 0.097 min^{-1}, being much higher than the melamine-derived *g*-CN (without S-doping) activated by acid using the same procedure ($k = 0.036\ min^{-1}$).

7.1.2 Emerging Chemicals

A CN with pyridinic N-vacancies (V_{N2C}-CN) was prepared by the further calcination of urea-derived *g*-CN at 550°C for 2 h in Ar and used for the photodegradation removal of various organic pollutants, acetaminophen (AP), bisphenol A (BPA), tetracycline hydrochloride (TC), RhB, and sulfamethoxazole (SMZ), by dispersing 20 mg catalyst in deionized water solution of these pollutants [15]. The degradation kinetics in different initial concentrations of AP are compared on the pristine *g*-CN and V_{N2C}-CN in Figure 7.7a, revealing a marked acceleration of AP-degradation. Even with an initial concentration as high as 30 mg/L, AP was dramatically degraded over the V_{N2C}-CN. In Figure 7.7b, degradation reaction constant k for various organic pollutants is compared on the pristine CN and V_{N2C}-CN, demonstrating that the introduction of N-vacancy (V_{N2C}) accelerates the degradation reaction of pollutants, about 2 times acceleration.

C-rich CN nanosheets exhibited a notably enhanced photocatalytic oxidation activity toward emerging organic pollutants, such as AP, levofloxacin (LEV), and

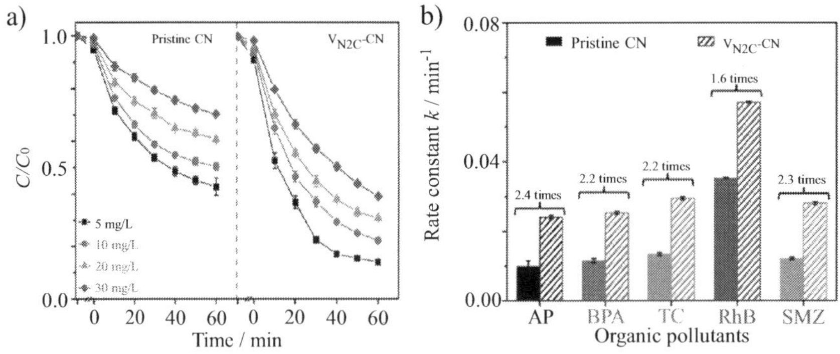

FIGURE 7.7 Photodegradation performances of the pyridinic N-deficient CN (V_{N2C}-CN) in comparison with the pristine CN: (a) the degradation kinetics in acetaminophen (AP) solutions with different initial concentrations and (b) degradation rate constant k for various organic pollutants [15].

methylparaben (MPB) [16]. The CN nanosheets were synthesized by mixing urea (0.5 mol) with L-cysteine ($C_3H_7NO_2S$) of 0.0, 0.1, 0.3, 0.5 and 1.0 mmol in deionized water, dehydrating at 100°C, and then heating at 180°C in an autoclave, and then annealing at 550°C for 2 h in air. The resultant CN is coded as C_X-CN, where X corresponds to L-cysteine used during the synthesis (0.0–1.0 mole). The photocatalytic oxidation activity of the resultant catalysts C_X-CN nanosheets to organic pollutants (AP and LEV) depends strongly on the level of carbon doping. The $C_{0.5}$-CN exhibited the fastest and 100% degradation within a short time under light irradiation, as shown in Figure 7.8a and b, respectively. The resultant catalysts exhibited similar but negligibly small adsorption capacities (less than 4.0%) for three pollutants, and the direct photolysis in the absence of catalysts could hardly be detected on these pollutants. As shown in Figure 7.8a and b on AP and LEV, the resultant C_x-CN catalysts show high pollutant removal efficiency under visible light irradiation, being much higher than those of $C_{0.0}$-CN (no C-doping), the reference g-CN and TiO_2 (P25). The $C_{0.5}$-CN catalyst exhibited excellent removal efficiencies for three pollutants under visible-light ($\lambda > 400$ nm) irradiation, reaching 90.3% for AP after only 10 min and 98.3% for LEV after only 8 min, but 99.3% for MPB after 90 min.

A defect-rich, porous CN nanosheets were deposited on a conductive FTO substrate, which has an extended optical absorption and has a photocatalytic activity for the degradation of benzylamine (BA) via oxidation [17]. The supramolecular aggregates (the CN precursors) were prepared by hydrothermal treatment of the mixture of melamine (M) with various amounts of HCl and then converted to a homogeneous paste by mixing with ethylene glycol (EG). It was pasted onto an FTO substrate and then calcined at 530°C in N_2 flow to achieve complete conversion of the precursor to CN. The resultant CN-films on a substrate exhibited a 1D configuration, being associated with abundant defects, as illustrated in Figure 7.9a, and were used as the photoanode of the electrochemical device, as illustrated in Figure 7.9b. The device delivered high photoelectrochemical activity for the oxidative degradation of BA as well as water splitting. The CN-photoanode produced a photocurrent of 183 μA/cm²

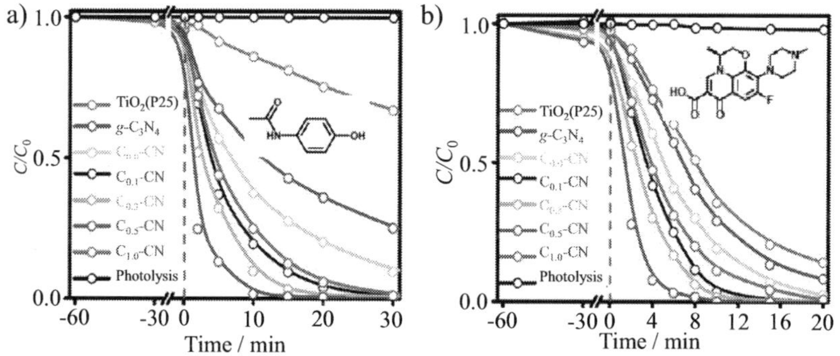

FIGURE 7.8 Photocatalytic activity to the degradation of the organic pollutants by C-rich CNs (C_X-CN) in comparison with TiO_2 (P-25) and the pristine g-CN: (a) for AP and (b) LEV. Molecular structure of the respective pollutant is shown in each figure as an inset [16].

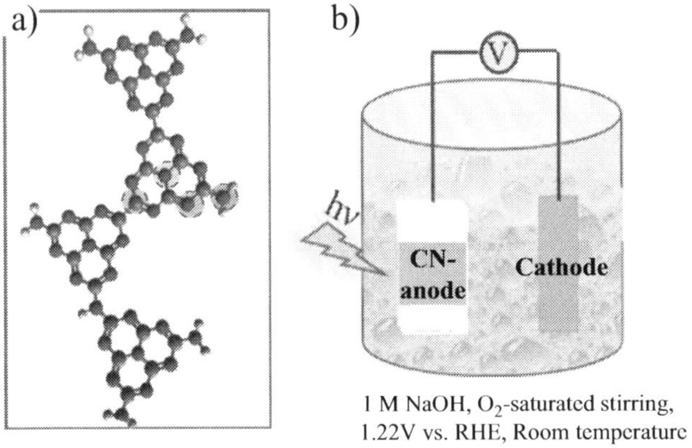

FIGURE 7.9 BA-degradation via oxidation over the defect-rich CN: (a) molecular structure of the defect-rich CN by highlighting the defects (broken black circles) and (b) the electrochemical device used and the experimental conditions [17].

at a potential of 1.22 V. In the presence of BA and O_2, the photocurrent reaches 200 μA/cm^2, signifying that benzylamine undergoes oxidation to benzaldehyde and imine (N-benzylideneaniline) at the photoanode more quickly than water-splitting. BA-degradation efficiency reached 56%.

Degradation performances of bisphenol were studied on two series of *g*-CN-derived photocatalysts, TCNs and CCNs, under the irradiation of blue LED light with a wavelength of 450 nm [18]. TCNs were synthesized from the mixtures of urea (10 g) with 1,000, 100, and 10 mg of trisodium citrate dihydrate ($C_6H_5Na_3O_7{\cdot}2H_2O$) in 20 mL of water and then calcined at 550°C and coded as TCN-1000, TCN-100, and TCN-10, respectively. CCNs are synthesized using urea and citric acid ($C_6H_8O_7$) in the same procedures as that for TCNs and coded as CCN-1000, CCN-100, and CCN-10. For CCN-10, the additional calcination at 350°C, 400°C, 500°C, 550°C, and 600°C was applied (coded as CCN-10-350, etc.). The degradation behaviors under irradiation for TCN- and CCN-series catalysts are shown together with the photocatalytic degradation rate for the respective catalyst curves, in comparison with the pristine *g*-CN in Figure 7.10a and b, respectively. The results exhibit a marked improvement in bisphenol-degradation activity by the addition of trisodium citrate and citric acid to the precursor urea. In particular, the addition of a small amount of citric acid (10 mg) was efficient. In Figure 7.10c, the effect of additional calcination at 350°C–600°C is shown on the CCN-10 (which had been prepared at 550°C). The catalyst applied the further calcination at 600°C, i.e., CCN-(550 + 600), exhibited the highest degradation rate (0.105 min^{-1}) and 100% degradation of bisphenol was achieved within 40 min of irradiation.

Mn and S co-doped *g*-CN (Mn, S-CN) was synthesized from a methanol solution of thiourea with $Mn(NO_3)_2$ by calcining at 520°C [19]. It exhibited an exceptional photocatalytic degradation removal efficiency for various pollutants, such as BPA, phenol (BP), 4-chlorophenol (4-CP), nitrobenzene (NBz), RhB, AP, and

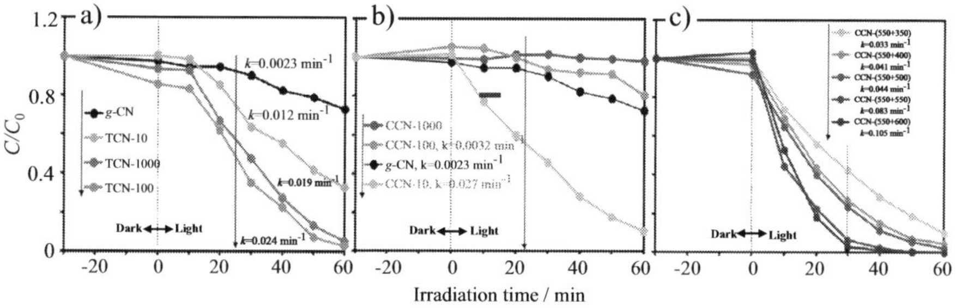

FIGURE 7.10 Degradation curves of bisphenol E in 5 ppm concentration: (a) TCN-series catalysts, (b) CCN-series catalysts in comparison with g-CN, and (c) effect of further calcination on CCN-10 [18].

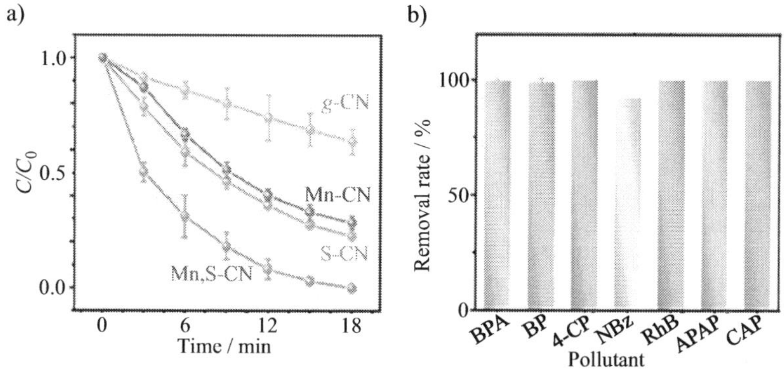

FIGURE 7.11 Mn and S co-doped CN-based photocatalysts (the pristine g-CN, S-CN, Mn-CN, and Mn, S-CN): (a) BPA degradation performances and (b) removal rate of various pollutants over the Mn, S-CN catalyst [19].

chloramphenicol (CAP). The Mn, S-CN was synthesized from a methanol solution of thiourea and $Mn(NO_3)_2$ by calcining at 520°C. For comparison, S-CN was prepared using the same procedure without $Mn(NO_3)_2$, and g-CN was prepared by replacing thiourea with urea by thermal decomposition. The degradation performances of the pollutants were measured by irradiation with simulated sunlight from an Xe lamp on a pollutant solution (10 mg/L). In Figure 7.11a, the BPA-degradation kinetics are compared with the resultant CN-based catalysts, revealing that the catalyst Mn, S-CN exhibits the highest degradation performance and the co-doping of Mn and S to g-CN is efficient. In Figure 7.11b, the removal efficiencies (%) of various pollutants over the Mn, S-CN catalyst are shown, demonstrating the excellent removal performances of Mn and S co-doped CN for various pollutants. The bandgaps for these CN-based catalysts were calculated to be in a range of 2.76–2.82 eV. On S-doped g-CN, the structure, electronic, and optical properties, as well as photocatalytic CO_2 reduction activity, were investigated and discussed [20].

g-CN is functionally not only a semiconductor but also a solid-state organic reagent having surface-terminating amino groups and lone-pair N atoms of the

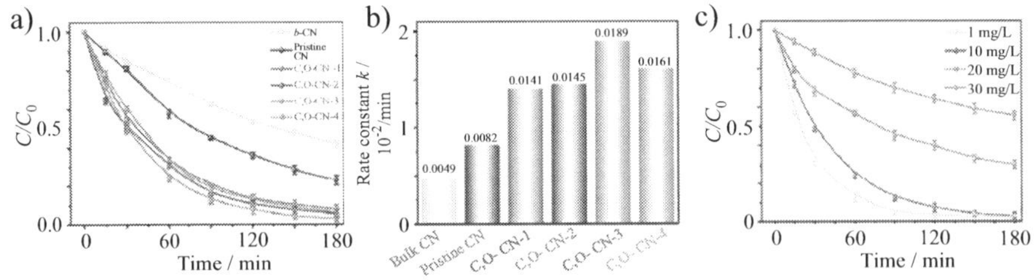

FIGURE 7.12 Photocatalytic degradation of oxytetracycline (OTC) by C- and O-co-doped CNs (C, O-CN-1–C, O-CN-4) in comparison with the pristine and bulk CNs: (a) degradation removal kinetics, (b) rate constant of OTC degradation, and (c) effect of initial concentration of OTC-degradation over C, O-CN-3 [22].

poly(heptazine) framework, which can activate gaseous O_2 with visible light for liquid-phase oxidation of various alcohols. The *mpg*-CN was shown to work as a metal-free photocatalyst to activate O_2 for the selective oxidation of benzyl alcohol (BA) under visible light irradiation [21]. The oxidation of alcohol to the corresponding aldehyde was conducted using a mixture of 10 mL solvent with BA (1 mmol) and *mpg*-CN (50 mg) at 100°C pressurizing with O_2 (8 bar) in a closed autoclave with a quartz window for introducing visible light from a 300 W Xe lamp with a 420 nm cut-off filter. Without the *mpg*-CN photocatalyst, BA was not converted to aldehyde, even under high-pressure oxygen for 3 h.

7.1.3 Antibiotics

C and O co-doped *g*-CNs, which were synthesized from the mixtures of cyanuric acid and melamine (1/1 by mol) with benzonitrile (50, 75, 100, and 125 μL) and coded as C-O-CN-1, -2, -3, and -4, respectively [22]. The resultant CN-based catalysts exhibited enhanced performance for the degradation removal of an antibiotic, oxytetracycline, OTC. The degradation curves of OTC by C- and O-co-doped CNs (C, O-CN-1–4) are shown in comparison with the pristine *g*-CN and the bulk CN (*b*-CN) in Figure 7.12a, revealing a marked acceleration of the degradation reaction over C, O-CNs, the curves being able to be approximated by the first-order formula, of which the rate constant k is shown in Figure 7.12b. The catalyst C, O-CN-3 exhibited the highest activity for OTC-degradation, as well as H_2-evolution. Figure 7.12c shows the effect of the initial OTC concentration on degradation kinetics, revealing that the degradation proceeded as fast as the decrease in the initial OTC concentration.

Photocatalytic degradation of TC hydrochloride in water over different *g*-CNs [23]. The *g*-CNs were synthesized from melamine and from the supramolecular aggregates derived from melamine and cyanuric acid in equal mass by calcining at 550°C for 4 h in air, respectively, and coded as CN and SCN, respectively. Additionally, *g*-CNs derived from supramolecular aggregates using different amounts of EDTA-2Na (0.1–0.5 g) were prepared under the same conditions (550°C for 4 h) and coded as E_X/SCN, where X corresponds to the amount of EDTA-2Na used. Photocatalytic degradation of TC was studied by dispersing 100 mg of catalyst in a 100 mL TC solution (50 mg/L) under visible

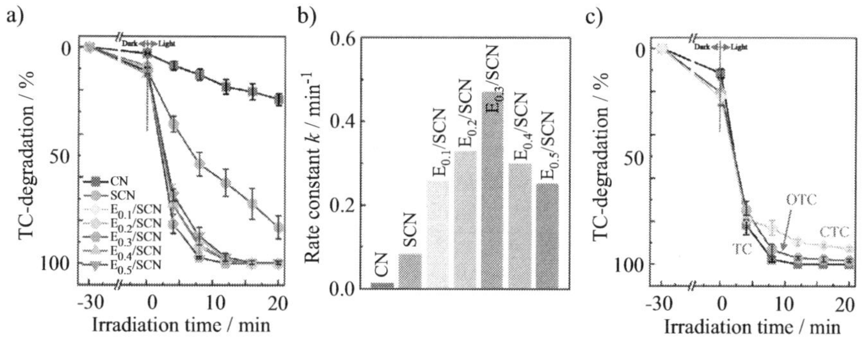

FIGURE 7.13 Photocatalytic degradation of TC (50 mg/L) over various g-CN catalysts under irradiation of light ($\lambda > 420$ nm) from 300 W Xe-lamp: (a) adsorption and degradation curves, (b) degradation rate constant k, and (c) adsorption and degradation curves over $E_{0.3}$/SCN for different antibiotics (TC, OTC, and CTC) [23].

light. TC-degradation kinetics and its rate constant k are compared on the resultant catalysts in Figure 7.13a and b, respectively. After reaching adsorption equilibrium in the dark (30 min), all E_X/SCN present much higher activity than the CN and SCN. The photocatalytic activity of the E_X/SCN first increased with the increase in the EDTA-2Na dosage from 0 to 0.3 g, and then gradually decreased as the dosage increased to 0.4 and 0.5 g. TC in the initial solution (50 mg/L) was almost completely removed by all E_X/SCN catalysts in 20 min, indicating that coupling of the morphological/structural regulation by the supramolecular self-assembly route synergistically improved photocatalytic behavior. As shown in Figure 7.13c, OTC and chlortetracycline (CTC) with an initial concentration of 50 mg/L were also effectively degraded by the $E_{0.3}$/SCN, which exhibit degradation efficiencies of 98.1% for OTC and 92.8% for CTC after 20 min visible light illumination. Single Ni atom (Ni1) and S co-doped g-CN, (Ni1, S-CN) exhibited a high TC degradation rate of 0.031 min^{-1}, being 6.0 times higher than the pristine CN [24].

N-defective g-CNs (V_N-CN-X) with different N-vacancy densities exhibited high photocatalytic activity for the degradation removal of ciprofloxacin (CIP) [25]. The introduction of V_N was performed by the heat treatment of the urea-derived g-CN by mixing with $NaBH_4$ at 300°C for different times, 0.5, 1, and 2 h, to get V_N-CN-X, where X corresponds to the time of $NaBH_4$-treatment. V_N-CN-1 exhibited the highest CIP-degradation performances, its efficiency reaching 98.8% and rate constant k 0.0461 min^{-1}, and V_N-CN-2 delivered very close performances (88.8% and 0.0129 min^{-1}), while the pristine CN shows 42.3% and 0.00479 min^{-1}, respectively.

7.1.4 Air Purification

g-CN and g-CN-based materials have been studied as candidates for air purifiers for the degradation removal of the oxygenated volatile organic compound (VOC). The NO_x-removal performances were studied on the chemically and thermally exfoliated g-CNs [26]. A melamine-derived g-CN was chemically exfoliated by treatment with H_2SO_4 for 1–5 h. The chemically exfoliated g-CN provided significantly high photocatalytic activity for NO_x-degradation in comparison with the pristine bulk and the thermally exfoliated *g*-CN. K-doped *g*-CN (K_X-CN) demonstrated high photocatalytic performance for formaldehyde (FA)-elimination [27]. K_X-CN photocatalysts were prepared by the calcination of a mixture of 10 g urea with the different amounts of KNO_3 (X=0.20, 0.33, 0.43, and 0.53 g) and calcination at 540°C for 2 h in a ceramic crucible. The $K_{0.43}$-CN catalyst exhibited the highest FA-degradation efficiency, exhibiting a marked FA-concentration reduction from 300 to 7 ppm during 32 min of visible light irradiation. A FA-reduction down to 99 ppm was observed over the $K_{0.20}$-CN catalyst. This FA-concentration reduction corresponded quantitatively to the CO_2-generation due to the FA oxidation to CO_2 (mineralization).

7.1.5 Arsenics

$C_3N_4/C_3N_{5.4}$ molecular heterojunctions were synthesized by mixing the melem with aminoguanidine hydrochloride (CH_6N_4·HCl) in different mass ratios (1/5, 2/52, 3/5, and 4/5) in 20 mL water, heating and stirring the solution until all the water evaporated and then being calcined at 500°C for 4 h in N_2 flow (a self-assembly and copolymerization process) [28]. The resultant $C_3N_4/C_3N_{5.4}$ molecular heterojunctions were coded as MJ-X, where X corresponds to aminoguanidine hydrochloride/melem ratio (1–4). On these catalysts, MJ-X, the performance of arsenic removal was studied by dispersing 50 mg of catalyst in a 10 mg/L sodium arsenite aqueous solution under O_2-bubbling for 6 h under light or dark. The experimental results revealed that the catalyst MJ-2 exhibited the highest As^{3+} and As^{5+} removal efficiency up to 44.40% and 77.48%, respectively, and that the heterojunction provides many adsorption sites. As^{5+} can be more effectively removed from solution than As^{3+}. The MJ-2 exhibited an excellent As^{3+} oxidation rate of 19.35 mg/g·h, which was much higher than C_3N_4 and $C_3N_{5.4}$.

7.1.6 Explosives

Tubular *g*-CN, which was prepared by liquid-phase exfoliation of the melamine-derived *g*-CN, exhibited a high electrochemical sensitivity for nitroaromatic explosives, including 2,4-dinitrotoluene, in water [29].

7.2 *g*-C_3N_4-BASED COMPOSITES

7.2.1 Organic Dyes

Composites of CN with tris(4-aminophenyl)amine (TP) were prepared by mixing with different TP/CN weight ratios of 5/95, 10/90, 20/80, 50/50, and 80/20 in absolute ethanol under sonication and then evaporating the solvent at 50°C [30]. The resultant powders were coded as TP_X/CN, where X corresponds to TP/CN weight ratio 5–80. Their photocatalytic activity for the removal of organic dye in a wastewater was evaluated by dispersing 10 mg TP_X/CN catalyst powder in 20 mL RhB solution (20 μmol/L concentration) under the irradiation of 50 W light (λ > 410 nm). Among the TP_X/CN-series catalysts, the TP_{20}/CN exhibited the highest RhB-degradation performance and adsorption capacity, being almost comparable to that of TP (about 30 mg/g and adsorption efficiencies of about 74%), while the pristine CN exhibited poor adsorptivity.

Two kinds of MoS_2/g-CN composites prepared by hydrothermal treatment (MoS_2/CN-HT) and by calcination (MoS_2/CN-Cal) delivered high degradation performances for RhB [31]. The changes in RhB-concentration due to the adsorption in the dark and due to the degradation under the simulated solar light by dispersing various catalysts are shown in Figure 7.14a. The composites can adsorb RhB, while the pristine g-CN and MoS_2 cannot. Under the solar light irradiation, the composites degrade RhB quickly, much faster than the pristine g-CN and MoS_2, the rate constant k for RhB-degradation being calculated to be 0.035 and 0.025 min^{-1} for the composite MoS_2/CN-HT and -Cal, respectively, much higher than 0.009 and 0.006 min^{-1} for the pristine g-CN and MoS_2. Figure 7.14b shows how much the RhB-degradation is reduced by adding 1 mmol p-benzoquinone (P-BQ), Isopropanol (IPA), and ethylenediamine-N,N,N′,N′-tetra-acetic acid disodium salt (EDTA-2Na), which are quench-trapping agents for $*O_2^-$, *OH and h^+, respectively. The degradation efficiency of MoS_2/CN-HT decreases from 91.2% to 37.1% by adding P-BQ after 1 h, while that of MoS_2/CN-Cal decreases from 82.4% to 35.2%. Similarly, by the addition of IPA and EDTA-2Na, the degradation efficiency of MoS_2/CN-HT decreases from 91.2% to 40.5% and 49.7%, respectively, while that of MoS_2/CN-Cal decreases from 82.4% to 41.2% and 48.0%, respectively. Since the quenching effect of IPA and EDTA-2Na appeared to be less significant in comparison with P-BQ, suggesting that $*O_2^-$ may play the most significant role in the photocatalytic degradation of RhB by the MoS_2/CN composites.

The self-composite $(B\text{-}CN)_X$/CNs delivered the high degradation performances of MO and phenol under visible light (>420 nm) irradiation [32]. The composite B-CN_X/CN was prepared by mixing urea (10 g) with different amounts of B-doped CN (B-CN) (X = 10, 25, and 50 mg) and being calcined at 550°C for 2 h. The B-CN was synthesized from the mixture of urea (10 g) with Ph_4BNa (6.0 mg) by calcination at 550°C. The resultant composites exhibited a high degradation performance for

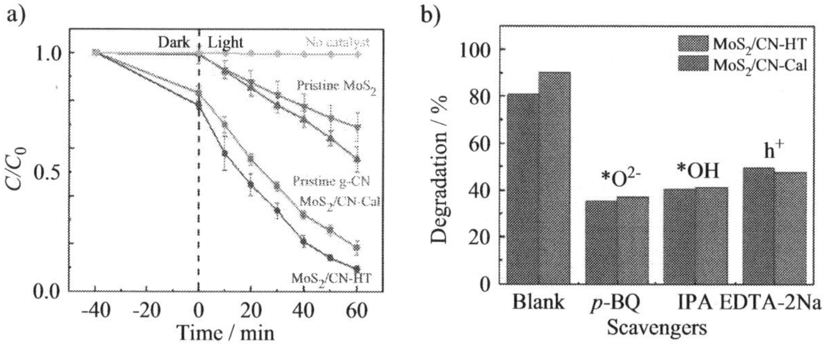

FIGURE 7.14 Photocatalytic degradation of Rhodamine B by two composites, MoS_2/CN-HT and MoS_2/CN-Cal: (a) adsorption and degradation kinetics of RhB and (b) the effect of quenching-agents under visible light irradiation [31].

both MO and phenol. After 2 h irradiation, approximately 93% of MO and 57% phenol were removed by the $(B\text{-}CN)_{25}$/CN composite, whereas less than 45% MO and less than 30% phenol were removed by the pristine CN and B-CN.

A ternary nanocomposite *g*-C_3N_4/TiO_2/carbon nanotube (CN/TiO_2/CNT) was prepared by hydrothermal treatment of H_2SO_4 solution of a mixture of titanium isopropoxide, the urea-derived *g*-CN and MWCNT at 150°C for 44 h [33]. A series of ternary nanocomposites CN/TiO_2/CNT were prepared using different amounts of CNT (0.4, 0.8, and 1.2 wt%), which are denoted as CN/TiO_2/CNT_X, where X corresponds to the CNT mixing ratio (0.4, 0.8, and 1.2). For comparison, the binary nanocomposite CN/TiO_2 and pristine TiO_2 were prepared under the same conditions without *g*-CN and/or CNT. The photocatalytic degradation (i.e., discoloration) of MB dye was investigated by dispersing 100 mg of catalyst in a 250 mL aqueous solution of MB (6 mg/L) under visible light from a 150 W halogen lamp with a UV cut-off filter. The MB-degradation curves and degradation rates are compared on the related catalysts, i.e., the pristine TiO_2 and *g*-CN, binary composite CN/TiO_2, and ternary composites CN/TiO_2/CNT_X, revealing the effectiveness of compositing with CNTs, particularly CN/TiO_2/$CNT_{0.8}$, giving a high k as 0.0390 min^{-1} in comparison with 0.0007 min^{-1} for the pristine TiO_2.

Lignin nanosphere-loaded nanotubular *g*-CNs, which were coded as L_X/*nt*-CNs, where X corresponds to the quantity of lignin nanospheres used during synthesis (1%, 5%, and 10%) worked effectively to remove a dye, brilliant black BN (BBBN, $C_{28}H_{17}N_5Na_4O_{14}S_4$) [34]. In Figure 7.15a, the degradation kinetics of BBBN over the resultant composites, including the pristine *nt*-CN, are compared, revealing that the composite L_5/*nt*-CN exhibited the highest performance. After 15 min of irradiation of a simulated solar light, L_5/*ut*-CN and the pristine *nt*-CN achieved BBBN removal efficiencies of 82% and 70%, respectively. After 60 min of light exposure, the removal efficiency of L_5/*nt*-CN for BBBN reached 99.3%. The degradation curves were approximated by a first-order relation and the rate constant k is shown as an inset in Figure 7.15a, revealing the L_5/*nt*-CN composite exhibited the highest k

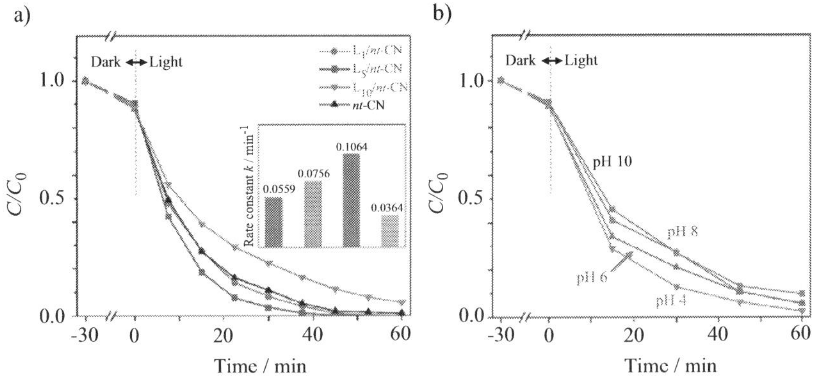

FIGURE 7.15 Lignin nanosphere-loaded nanotubular *g*-CNs (Lx/*nt*-CN composites): (a) BBBN-degradation kinetics over the composites and the rate constant k for the respective composite (inset), and (b) effect of pH of BBBN-solutions using L_5/*nt*-CN catalyst [34].

(0.1064 min^{-1}). In Figure 7.15b, the effect of the pH of the BBBN-solution is shown on L_5/*nt*-CN composite. Under acidic conditions, the composite exhibited a high degradation efficiency toward BBBN. Moreover, with the increase in pH value, the removal efficiency of BBBN increased slightly.

7.2.2 Antibiotics

The self-composite AOP-CN, which was composed of a ternary homojunction between A-CN (amino-rich *g*-CN), O-CN (oxidized *g*-CN), and P-CN (phosphorylated *g*-CN), delivered a high performance for the photodegradation of tetracycline (TC) [35]. AOP-CN composite hydrogel exhibited the highest performance in comparison with the constituent catalysts, such as A-CN, O-CN, including the bulk AOP-CN, of which the removal efficiency was calculated to be 31% in 40 min. These results demonstrated how effective the hydrogelation of *g*-CN is for the degradation of TC.

The CoPc-loaded *mpg*-CN nanotubes containing N-vacancies were synthesized and their photocatalytic activity for the degradation of sulfamethoxazole (SMX) under visible light irradiation with the assistance of peroxymonosulfate (PMS) was investigated [36]. The catalysts were prepared by calcination of the mixtures of urea (14.4 g), melamine (1.2 g) and different amounts of Co(II) phthalocyanine (CoPc) (5, 10, 15, and 20 mg) at 550°C and the resulting catalysts were coded as $(CoPc)_X$/CN, where X corresponds to the amount of CoPc used in the synthesis of the catalyst. The photocatalytic degradation performances for SMX were measured by dispersing the catalyst powder in a SMX aqueous solution (2 mg/L) by adding 0.2 mmol PMS. The results are shown in Figure 7.16a, exhibiting that the relative concentration of SMX decreasing the faster with increasing CoPc-content in the composite catalysts, reaching about 100% degradation by the resultant composites, $(CoPc)_{10}$/CN, $(CoPc)_{15}$/CN and $(CoPc)_{20}$/CN, within 40 min irradiation of visible light, whereas the pristine CN gave only 36% degradation. Particularly, the catalyst $(CoPc)_{20}$/CN

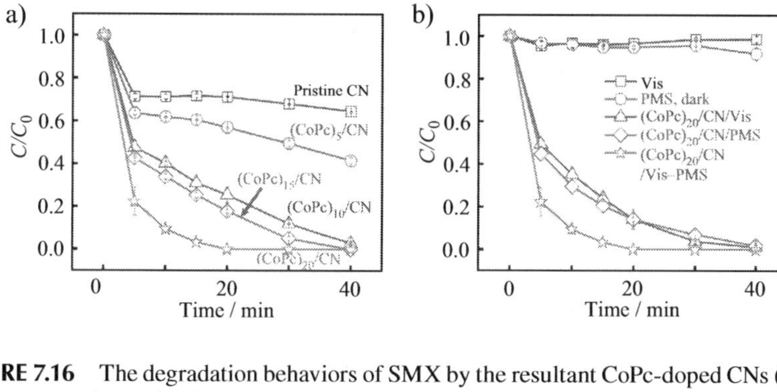

FIGURE 7.16 The degradation behaviors of SMX by the resultant CoPc-doped CNs ($CoPc_x$/CN) in comparison with the pristine CN: (a) by $CoPc_x$/CN catalysts with different Co-doping and (b) by the $CoPc_{20}$/CN catalyst under different conditions, co-existence of PMS and/or visible light irradiation. Concentrations of SMX were 2 mg/L, that of the catalyst is 0.2 g/L, and that of PMS is 0.2 mmol [36].

gave an excellent performance, SMX being completely degraded within 20 min irradiation. The rate constant *k* for the SMX-degradation over the pristine CN and $(CoPc)_5$/CN, $(CoPc)_{10}$/CN, $(CoPc)_{15}$/CN, and $(CoPc)_{20}$/CN was calculated to be 0.007, 0.017, 0.079, 0.091, and 0.222 min^{-1}, respectively. In Figure 7.16b, the degradation behaviors of SMX in different systems, by dispersing the catalyst $(CoPc)_{20}$/CN with PMS under visible light ($(CoPc)_{20}$/CN/Vis-PMS system), $(CoPc)_{20}$/CN with PMS in the dark ($CoPc_{20}$/CN/PMS), $(CoPc)_{20}$/CN without PMS under visible light ($(CoPc)_{20}$/CN/Vis), in comparison with only visible light irradiation without the catalyst (Vis) and just adding PMS without the catalyst in the dark (PMS), are compared. The figure reveals that the existence of the photocatalyst $(CoPc)_{20}$/CN under the visible light irradiation, together with the assistance of PMS, is essential to achieve 100% degradation of PMX.

Fe phthalocyanine (FePc) was also loaded onto O-doped porous CN (O-CN), forming a photocatalyst working in an assisted PMS activation for the removal of TC [37]. FePc-loading onto the O-CN was carried out by mixing the 200 mg O-CN with different amounts of FePc (1, 10, and 15 mg) in DMF and following sealing for 24 h and washing. The photocatalytic activity for the TC-removal was studied on the resultant composite $(FePc)_X$/O-CN by dispersing 0.1 g/L catalyst powder in a water containing 25 mg/L TC with 0.5 mmol/L PMS under the light from a 300 W Xe lamp at approximately 25°C. The pristine CN with PMS-assistance achieves a TC-degradation of only 46.7% under visible light for 30 min. The use of O-doped CN (O-CN) improves the degradation efficiency to 62.7%. FePc (without CN) can rapidly degrade TC within 3 min, but the degradation efficiency is limited to approximately 70%. Even $(FePc)_1$/O-CN, which has relatively low FePc-loading, can degrade 87.1% TC within 30 min. Under the sole action of PMS, the TC-degradation reaches 31%, owing to the strong oxidation capability of PMS. $(FePc)_{10}$/O-CN demonstrates greater PMS activation ability than O-CN under visible light, resulting in an increase in TC-degradation from 62.7% to 100%. The co-existence of Cl^-, CO_3^-, HCO_3^{2-}, and PO_4^{3-} in the solution had minimal effect on the degradation of TC, while humic acid had a slight inhibitory effect on TC-removal. Loading of an iron porphyrin MOF (FeP/MOF) onto protonated *g*-CN (FeP/MOF/H-CN composite) also enhanced the photodegradation of TC and H_2-evolution [38].

B-doped N-deficient *g*-CN (V_N-B_X-CN) was synthesized using different mass ratios of $NaBH_4$ to the matrix *g*-CN (*X*=0.2, 0.4, 0.6 and 0.8 g to 0.8g *g*-CN)) at different temperatures (*Y*=300°C, 350°C, 400°C, 450°C, and 500°C) in N_2 for 1 h, and were coded as V_N-B_X-CN-*Y* [39]. $Cs_3Bi_2Br_9$ perovskite nanocrystals were loaded onto the composites in a dimethyl sulfoxide solution. $Cs_3Bi_2Br_9$ was employed instead of $CsPbX_3$ (X=Cl, Br, I) perovskite nanocrystal to get a Pb-free photocatalyst. $Cs_3Bi_2Br_9$/V_N-$B_{0.6}$-CN-450 exhibited the highest degradation efficiency of 98.62% for chloroquine phosphate (CQ) within 60 min and a high oxidation ability (>82.19%). This composite not only possesses a high degradation rate for a variety of antibiotics, but also exhibits excellent structural stability and reusability. The Bi_2O_3/BN/Co_3O_4 ternary composites delivered a high degradation removal performance against norfloxacin under visible light irradiation [40], reaching 98% degradation in 180 min using 5 wt% catalyst.

7.2.3 Water Purification

g-CN loaded onto TiO_2 (CN/TiO_2 composite) was studied for water purification [41]. CN/TiO_2 composite was prepared by mixing TiO_2 (anatase type), which was prepared from titanium isopropoxide, with different amounts of CN (2%, 4% and 8% to TiO_2) by grinding and following calcination at 600°C–1,000°C for 2 h, the resultant composite was coded as $(CN)_X/TiO_2$, where X corresponds to the CN mixed with TiO_2 (2, 4 and 8). The degradation rate under UV and visible light irradiation is summarized with particle size of anatase and rutile phases of TiO_2 and the bandgap for the composites is shown on the composites together with the composite preparation conditions (calcination temperature and TiO_2-content) with the bandgap of the composite. The effectiveness of the composite catalyst is shown in Table 7.1 together with the preparation conditions, etc. The degradation rate constant k of the composites $(CN)_X/TiO_2$ for Rohdamine 6G in water under UV and visible light irradiation are shown with the preparation conditions and the bandgap of the composites in Table 7.1. The photocatalytic activity of the resultant composites was estimated by measuring the degradation rate of Rohdamine 6G, which was increased under UV irradiation, revealing the effectiveness of the composite catalyst. Calcination at 600°C partially transformed one of the components, i.e., TiO_2, from the photoactive anatase phase to the inactive rutile phase. The photocatalytic activity of the composite increased with increasing CN-content up to 4%, but 8% addition of CN could not compensate for the decrease in the photoactivity due to the phase transformation of TiO_2.

A Sm-doped $GdFeO_3$/g-CN with a heterostructure-based Z-scheme photocatalyst was fabricated for the effective removal of the hazardous environmental pollutants, such as Malachite Green (MG) dye in water. [42]. The composites of g-CN with anthraquinone compounds (AQX/CN) exhibited a high degradation rate of sulfadiazine (SDZ, a bacterium-infecting antibiotic) under conditions of high chloride ion (Cl^-) concentration [43].

7.2.4 Air Purification

The high adsorption performances for toxic gas molecules, including CO, NO, NO_2, SO_2, and H_2S of the pristine and metal-doped g-CNs (Cu, V, and Fe), were theoretically predicted [44]. The high specific surface area, mechanical properties, thermal stability, and flexible surface chemistry of 2D g-CN are advantageous in many environmental applications, and the pore-like structure provides additional sites for gas adsorption. An atomic lattice composed of s-triazine or tri-triazine groups in crystalline g-CN provides a porous diameter of 0.34 nm, which is highly desirable to adsorb these gas molecules.

The g-CN powder was functionalized with polyacrylamide (Acryl) dispersed highly in an EG solution [45]. It was further cross-linked with citric acid (Cit) molecules to obtain an ink with a high viscosity. The resultant inks containing g-CN with different combinations and ratios of CN/Acryl, Acryl/Cit, and EG/Acryl were cast onto the various substrates, fluorine-doped tin oxide (FTO), plexiglass, metal-foil, wood, etc., and cured at 100°C to obtain the photocatalyst composite films with a thickness of about 212 μm. The catalyst films were tightly fixed on each substrate and

TABLE 7.1
$(CN)_X/TiO_2$ Composites in Comparison with the Pristine TiO_2 and *g*-CN [41]

Catalyst	Calcination Temperature (°C)	Content of *g*-CN (%)	Rhodamine 6G Degradation Rate *k* (min^{-1})		Anatase-Content (%)	Particle Size (nm)		Bandgap (eV)
			Under UV	Under Visible Light		Anatase	Rutile	
Pristine *g*-CN		100	0.130	0.012	–	–	–	2.82
Pristine TiO_2		0	0.034	0.005	100	23.9	0	3.15
$(CN)_2/TiO_2$	600	2	0.141	0.009	93.3	21.7	28.5	3.06
$(CN)_4/TiO_2$		4	0.187	0.009	89.8	22.7	46.8	3.09
$(CN)_8/TiO_2$		8	0.103	0.011	92.7	22.2	36.8	3.11

exhibited excellent photocatalytic performance under visible light for the removal of either polar or nonpolar VOCs, including gaseous acetaldehyde (Act) and toluene (Tol) with high concentrations,150 and 50 ppmv, respectively.

The removal of ethylene (C_2H_4) from the environment for the storage of fruits and vegetables is essential for extending their shelf-lives. For this purpose, the composite Ag_3PO_4-loaded *g*-CN nanosheets with *S*-scheme heterojunction were shown to be promising [46]. The melamine-derived *g*-CN powder was protonated by stirring in a 1 M HCl solution for 6 h. The protonated CN (0.2 g) was dispersed in deionized (DI) water with Na_2HPO_4 by stirring for 6 h to achieve full adsorption of HPO_4^{2-} onto the surface of the CN nanosheets, and then $AgNO_3$ solution (0.1 g/L) was dropped in the dark and finally obtained Ag_3PO_4 (APO) nanoparticles-loaded CN, which were coded as APO_X/CN composite, where X corresponds to the APO mass ratio (10%–70%). The photocatalytic activity of the pristine *g*-CN to degrade C_2H_4 is negligibly small, and only 10% of C_2H_4 decomposed during 300 min irradiation. The activity of Ag_3PO_4 nanoparticles was relatively better than the pristine *g*-CN, degrading 58% of C_2H_4 during 300 min. However, the composites of these two components, i.e., APO_X/CN, resulted in high performance for C_2H_4-degradation, the composite APO_{30}/CN degraded C_2H_4 completely within 120 min.

A new kind of S-scheme heterojunction photocatalyst was proposed by sandwiching *g*-CN between two semiconducting metal oxides, TiO_2 and ZnO, forming a ZnO/CN/TiO_2 2D composite [47], which contains two kinds of S-heterojunction as illustrated in Figure 7.17a. It was prepared by dispersing the nanoparticles of TiO_2 (1 g) and ZnO (0.2 g) with *g*-CN nanosheets (0.12 g) derived from N-acetyl D-glucosamine (GA) in 50 mL IPA under sonication and following slow evaporation to obtain Z(0.2)/CN(0.12)/T(1.0) composite (coded as ZTG-1). For comparison, the additional ternary composites Z(0.8)/CN(0.12)/T(1.0) (coded as ZTG-2) and binary composite TiO_2/CN and ZnO/CN were also prepared. In Figure 7.17b, degradation efficiencies of different VOCs (acetaldehyde, ethyl acetate, toluene, and benzaldehyde) over the resultant ternary and binary catalysts are compared under UV irradiation. Ternary composite, particularly ZTG-2, exhibited higher degradation efficiency than binary composites.

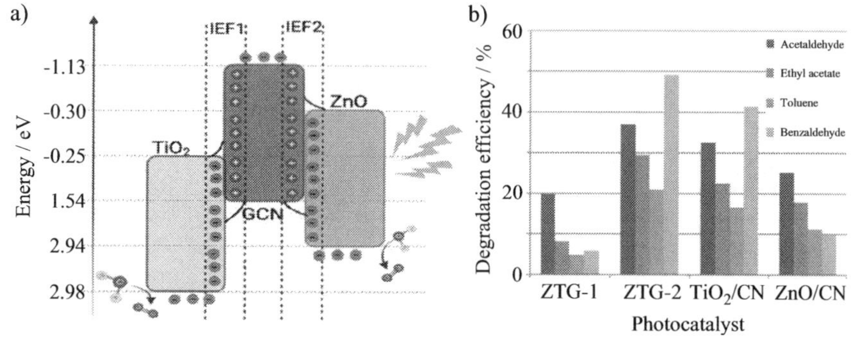

FIGURE 7.17 Ternary sandwich composites with two heterojunctions, ZnO/*g*-CN/TiO_2 composites: (a) illustration of sandwich-type two heterojunctions and (b) VOC-degradation efficiency of the ternary and binary composites [47].

The loading of 7,7,8,8-tetracyanoquinodimethane (TCNQ) onto *g*-CN using well-dispersed *g*-CN particles in a TCNQ-DMF (5 mg/mL) aqueous solution under sonication [48]. The phenol degradation performances of the TCNQ-loaded CNs (TCNQ/CNs) were studied by measuring the dependences of degradation rate constants on TCNQ-loading fractions and wavelength of the irradiation light. Only 10 wt%-loading of TCNQ onto the CN was effective for the acceleration of the phenol photodegradation. The photocatalytic activity of TCNQ/CN composite was also experimentally confirmed for 2,4-dichlorophenol and BPA.

7.2.5 Antipyretics

A composite of the melamine-derived *g*-CN with a COF was synthesized by microwave-assisted solvothermal treatment of the mixtures of *g*-CN with 5–25 wt% COF in DMF at 120°C for 30 min (COF/CN composites) [49]. The degradation of acetaminophen (ACE, an antipyretic) in water was studied on these composites with heterojunctions. There was a high and marked increase in absorbance in the UV region of UV-vis spectra for the composite with increasing COF content. The photocatalytic degradation of 5 mg/L of ACE was markedly accelerated by the addition of 0.5 g/L of the COF/CN composite. The maximum pseudo-first order rate constant k was calculated as 2.28 h^{-1} for the composite 20 wt% COF-loaded CN.

7.2.6 Bacteria

At a basic pH as 10, Cu^{2+} yielded coordination compounds with CN, as (Cu1)-accommodated at the intrinsic V_{6N} in *g*-CN framework [50]. It exhibited photocatalytic potential to generate superoxide anion radicals for disinfection and degradation of bacteria. The photo-disinfection kinetics of *E. coli* 25922 and deoxycholate (DXC) were measured. The (Cu1)-CN exhibited high photoactivity for disinfection and degradation of bacteria.

The composites of *g*-CN with a bacterial cellulose (BC) aerogel disk (4.9 cm diameter and 0.5 cm thickness and bulk density of 2.6–11.5 kg/m^3) were applied to an air filter to improve indoor air pollution caused by airborne bacteria, FM and particulate matter (PM) [51]. The resultant composites were coded as CN_X/BC, where X corresponds to the CN-content of 0, 34.1, 47.8, and 59.3 wt%, respectively. In Figure 7.18a, UV-vis absorption spectra are shown, revealing that the resultant composites, except CN0/BC, i.e., the pristine BC, are photoactive. In Figure 7.18b, the efficiencies for the PMs (PM-2.5 and PM-10) and pressure drops being associated with filtration are shown for respective composites. The resultant composites CN_X/BC exhibit almost 100% removal for both sizes of particulates, PM-2.5 and PM-10 mainly due to the entangled and porous BC, the surface hydroxyl groups can serve as adsorption sites for PMs. Figure 7.18c shows the kinetics of FM degradation removal over CN_0/BC (i.e., BC itself) and the $CN_{47.8}$/BC composite in the dark and under solar light irradiation. A small decrease in C/C_0 is due to physical adsorption by BC, while the $CN_{47.8}$/BC composite exhibits very rapid FM degradation under solar light irradiation due to photocatalytic degradation by *g*-CN loaded on the BC.

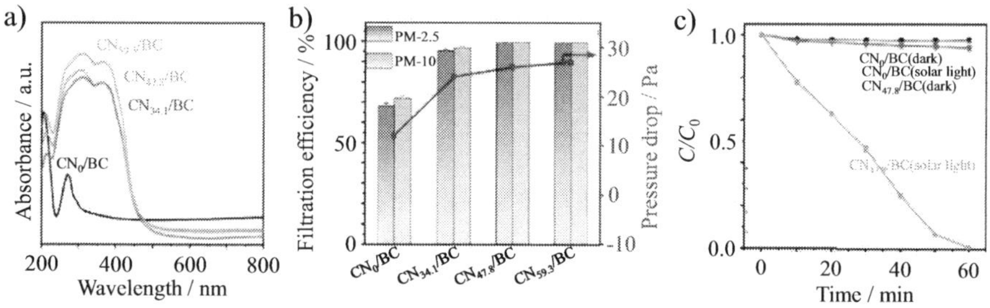

FIGURE 7.18 Air filtering by CN_x/BC composites: (a) UV-vis absorbance spectra, (b) filtration efficiencies for PM-2.5 and PM-10, and (c) formaldehyde removal performances in the dark and in solar light [51].

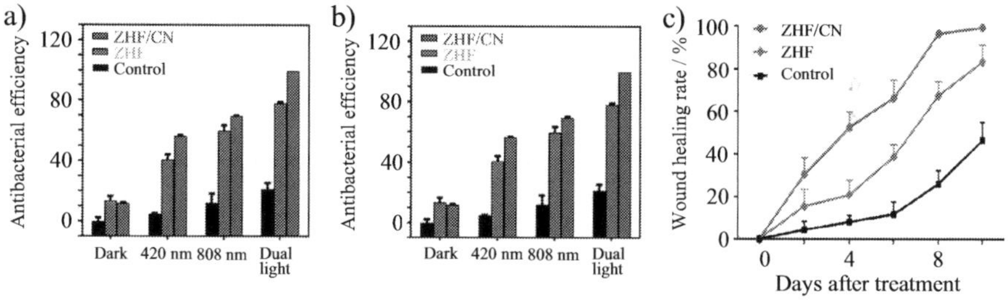

FIGURE 7.19 Antibacterial activity and wound healing rate of ZHF/CN composite and ZHF: antibacterial efficiency under different light irradiation conditions for (a) *S. aureus,* (b) *E. coli*, and (c) wound healing efficiency [52].

A g-CN-based nanocomposite with zinc hexacyanoferrate ($Zn_3[Fe(CN)_6]$, ZHF) nanospheres coded as ZHF/CN exhibited an enhanced antibacterial photocatalytic activity under dual-light irradiation (420 + 808 nm) [52]. In Figure 7.19a and b, antibacterial effect using a phosphate-buffered saline solution (PBS) of ZHF and ZHF/CN composite with 200 μg/mL concentration are compared in the dark and under 420, 808 nm and dual (420 + 808 nm) laser irradiation (power of 1.0 W/cm^2) over a period of 10 min on two bacteria, *E. coli* and *S. aureus* grown in Luria-Bertani (LB) medium at 37°C, respectively. Both ZHF and ZHF/CN exhibited antibacterial activity even in the dark (without light). However, an antibacterial effect under any light condition without catalysts was ineffective. Under light irradiation conditions, in contrast, the composite ZHF/CN exhibited higher antibacterial activity than ZHF itself, and dual light was effective for both bacteria, antibacterial efficiencies of ZHF/CN and ZHF against *S. aureus* reaching 99.87% and 78.12%, and against *E. coli* 99.98% and 85.06%, respectively. In Figure 7.19c, in vivo wound healing under dual light irradiation by the treatment of ZHF/CN and ZHF solution is shown by plotting the relative decrease in wound area against days after wounding, in comparison with the control experiment

(PBS solution without catalyst particles) on the basis of the results on 12 male mice. The wound area (8 mm diameter) was created on the back of a mouse by shaving and subsequent dropping of *S. aureus* suspension (10 µL) and then its healing treatment with the ZHF/CN and ZHF solution (200 µg/mL concentration) was applied for 15 min. The results (Figure 7.19c) demonstrated that both the ZHF/CN and ZHF showed enhanced healing effect, healing rate with the ZHF/CN reaching 99.22%, much better than that of the ZHF (83.30%) after 10 days.

7.3 CONCLUDING REMARKS

In this chapter, the degradation performances of various pollutants are focused, but the mentioned pollutants are so wide range from organic dyes, various emerging chemicals, various antibiotics, to bacteria. However, it must be pointed out that *g*-CN-related materials could convert a chemical to another, for example, the Fe-loaded *g*-CN could oxidize benzene to phenol under visible light irradiation [53]. In Chapter 9, the application of *g*-CN-related materials to the biomedical field is proposed by presenting some of the experimental results mentioned in this chapter as one of the prospects of the *g*-CN.

REFERENCES

1. Bhanderi D, Lakhani P, Modi CK, Graphitic carbon nitride (*g*-C_3N_4) as an emerging photocatalyst for sustainable environmental applications: A comprehensive review. *RSC Sustain* 2024; 2: 265.
2. Banu A, Sinha B, Sikdar S, Synthesis of polymeric 2D-graphitic carbon nitride (*g*-C_3N_4) nanosheets for sustainable photodegradation of organic pollutants. *Heliyon* 2024; 10: e33354.
3. Ilakiya S, Keerthana M, Ponpandian N, et al., Fabrication of graphitic carbon nitride synthesized via pyrolysis for environmental remediation: A detailed experimental analysis with different parametric optimizations. *New J Chem* 2024; 48: 7921–39.
4. Jun Y-S, Lee EZ, Wang X, et al., From melamine-cyanuric acid supramolecular aggregates to carbon nitride hollow spheres. *Adv Funct Mater* 2013; 23: 3661–7.
5. Simanek EE, Mammen M, Gordon DM, et al., Design and synthesis of hydrogen-bonded aggregates, theory and computation applied to three systems based on the cyanuric acid-melamine lattice. *Tetrahedron* 1995; 51: 607.
6. Shalom M, Inal S, Fettkenhauer C, et al., Improving carbon nitride photocatalysis by supramolecular preorganization of monomers. *J Am Chem Soc* 2013; 135: 7118–21.
7. Cui Y, Ding Z, Liu P, et al., Metal-free activation of H_2O_2 by *g*-C_3N_4 under visible light irradiation for the degradation of organic pollutants. *Phys Chem Chem Phys* 2012; 14: 1455–62.
8. Hu S, Li F, Fan Z, et al., Band gap-tunable potassium doped graphitic carbon nitride with enhanced mineralization ability. *Dalt Trans* 2015; 44: 1084–92.
9. Zhao J, Ma L, Wang H, et al., Novel band gap-tunable K–Na co-doped graphitic carbon nitride prepared by molten salt method. *Appl Surf Sci* 2015; 332: 625–30.
10. Yan SC, Li ZS, Zou ZG, Photodegradation of Rhodamine B and methyl orange over boron-doped *g*-C_3N_4 under visible light irradiation. *Langmuir* 2010; 26: 3894–901.
11. Gao X, Li QY, Wang YL, et al., A facile soft-hard template cooperative organization approach for mesoporous *g*-C_3N_4 with high photocatalytic performance. *Appl Surf Sci* 2024; 657: 159574.

12. Durairaj S, Bhatt H, Murugesan P, et al., Unveiling the mechanistic aspects of solar-driven hematein remediation using biocompatible metal-free carbon nitride nanosheets. *J Phys Chem C* 2024; 28: 11612–25.
13. Duhan J, Bisht A, Kumar H, et al., Graphitic nitride co-doped with sodium and sulphur as fluorescent and UV-Vis probe for quercetin detection. *Surf Interfaces* 2024; 52: 104927.
14. Benedoue SA, Ndi JN, Ge M, Tian Z, Dantio Nguela CB, Controllable acid activation of sulfur-doped carbon nitride enables enhancement in photocatalytic Indigo Carmine degradation performance. *ACS Sustain Chem* 2024; 12(45): 16809–16.
15. Yue J, Yang H, Liu C, et al., Unraveling the pyridinic nitrogen vacancy in carbon nitride for photo-self Fenton-like purification of organic contaminants. *J Colloid Interface Sci* 2024; 673: 475–85.
16. Zhang X, Liu Y, Ren M, et al., Precise carbon doping regulation of porous graphitic carbon nitride nanosheets enables elevated photocatalytic oxidation performance towards emerging organic pollutants. *Chem Eng J* 2022; 433: 134551.
17. Mondal S, Mark G, Tashakory A, et al., Porous carbon nitride rods as an efficient photoanode for water splitting and benzylamine oxidation. *J Mater Chem A* 2024; 12: 11502–10.
18. Ohnishi A, Furukawa M, Tateishi I, et al., Extension of π-conjugated aromatic structure in g-C_3N_4 nanosheets and their applications into bisphenol E decomposition. *J Solid State Electrochem* 2024; 28: 4527–49.
19. Sui C, Nie Z, Xie X, et al., Mn/S diatomic sites in C_3N_4 to enhance O_2 activation for photocatalytic elimination of emerging pollutants. *J Environ Sci* 2024; 149: 512–3.
20. Wang Y, Tian Y, Yan L, et al., DFT Study on sulfur-doped g-C_3N_4 nanosheets as a photocatalyst for CO_2 reduction reaction. *J Phys Chem C* 2018; 122(14): 7712–9.
21. Su F, Mathew SC, Lipner G, et al., mpg-C_3N_4-catalyzed selective oxidation of alcohols using O_2 and visible light. *J Am Chem Soc* 2010; 132: 16299–301.
22. Liu D, Zhao C, Chen M, et al., Enhanced visible light photocatalytic performance of carbon and oxygen co-doped carbon nitride with a three-dimensional structure: Performance and mechanism study. *J Colloid Interfaces Sci* 2024; 665: 452–64.
23. Shao S, Liu X, Wang R, et al., Morphologic and microstructural modulation of graphitic carbon nitride through EDTA-2Na mediated supramolecular self-assembly route: Enhanced visible-light-driven photocatalytic activity for antibiotic degradation. *Appl Surf Sci* 2024; 669: 160501.
24. Xu JX, Chen YF, Chen M, et al., *In situ* growth strategy synthesis of single-atom nickel/sulfur co-doped g-C_3N_4 for efficient photocatalytic tetracycline degradation and CO_2 reduction. *Chem Eng J* 2022; 442: 136208.
25. Liu Y, Chen X, Kamali M, et al., Unraveling the presence and positions of nitrogen defects in defective g-C_3N_4 for improved organic photocatalytic degradation: Insights from experiments and theoretical calculations. *Adv Funct Mater* 2024; 34(40): 2405741.
26. Papailias I, Todorova N, Giannakopoulou T, et al., Chemical vs thermal exfoliation of g-C_3N_4 for NO_x removal under visible light irradiation. *Appl Catal B* 2018; 239: 16–26.
27. Song S, Lu C, Wu X, et al., Strong base g-C_3N_4 with perfect structure for photocatalytically eliminating formaldehyde under visible-light irradiation. *Appl Catal B* 2018; 227: 145–52.
28. Song H, Yang Y, Chen P, et al., Dual regulation platform in $C_3N_4/C_3N_{5.4}$ molecular heterojunctions for simultaneous photocatalytic oxidative and adsorption of arsenic. *Sep Purif Technol* 2024; 335: 126200.
29. Udayan APM, Kaur B, Sensitive and fast electrochemical recognition of nitroaromatic explosives in liquid phase using ultrathin graphitic carbon nitride nanotubes. *Mater Chem Phys* 2024; 319: 129380.

30. Irshad UB, Akhtar Z, Bucher G, et al., Adsorption and photocatalytic properties of tris (4-aminophenyl) amine-based polyimide/graphitic carbon nitride composites for organic dye removal. *ACS Appl Polymer Mater* 2024; 6(6): 3390–401.
31. Zhang K, Yuan H, Xu F, et al., Rational fabrication of MoS_2/*g*-C_3N_4 heterostructures for efficient photocatalytic degradation of Rhodamine B. *Chem Select* 2024; 9(30): e202401034.
32. Li H, Liu Y, Gao X, et al., Facile synthesis and enhanced visible-light photocatalysis of graphitic carbon nitride composite semiconductors. *ChemSusChem* 2015; 8: 1189–96.
33. Chaudhary D, Vankar VD, Khare N, Noble metal-free *g*-C_3N_4/TiO_2/CNT ternary nanocomposite with enhanced photocatalytic performance under visible-light irradiation via multi-step charge transfer process. *Solar Energy* 2017; 158: 132–9.
34. Zhu C, Xiao X, Wang X, et al., Lignin-modified graphitic carbon nitride nanotubes for photocatalytic H_2O_2 production and degradation of brilliant black BN. *Inter J Biol Macromol* 2024; 267(2): 131533.
35. Balakrishnan A, Chinthala M, Kumar A, et al., Construction of dual Z-scheme ternary carbon nitride homojunction pectin microspheres as a multifunctional photocatalyst for tetracycline degradation, H_2O_2 production, and N_2 fixation. *Chem Eng J* 2024; 496: 153899.
36. Dou L, Huang W, Huang Y, et al., Carbon-rich *g*-C_3N_4 nanotubes with highly dispersed cobalt as a catalyst to degrade sulfamethoxazole in the vis-PMS system: Nonradical oxidation role. *Ind Eng Chem Res* 2024; 63(13): 5697–709.
37. Song Y, Zhang Y, Wu Y, et al., Construction of a photo-assisted peroxymonosulfate activation system comprised of iron phthalocyanine/O-doped carbon nitride for efficient tetracycline removal. *J Environ Chem Eng* 2024; 12(2): 112469.
38. Zhu W, Qin L, Ou Z, et al., Protonated carbon nitride cooperated with iron porphyrin metal–organic framework for boosting photocatalytic hydrogen production and degradation of tetracycline. *Fuel* 2024; 376: 132665.
39. Heng S, Xia M, Lv L, et al., Rational design of efficient boron-doped and nitrogen-deficient *g*-C_3N_4/lead-free $Cs_3Bi_2Br_9$ perovskite nanocrystals Z-scheme heterojunction by optimised surface-active site and interfacial charge transfer. *Sepa Purif Technol* 2025; 354(6): 129260.
40. Wang N, Wang W, Qi D, et al., Development of efficient and economic Bi_2O_3/BN/Co_3O_4 composite photocatalyst: Degradation mechanism, pathway and toxicity study of norfloxacin. *Chemosphere* 2024; 352: 141481.
41. Fagan R, McCormack DE, Hinder SJ, et al., Photocatalytic properties of *g*-C_3N_4-TiO_2 heterojunctions under UV and visible light conditions. *Materials* 2016; 9: 286.
42. Bibi F, Jamshaid M, Al-onazi WA, et al., Design of Sm-doped $GdFeO_3$/*g*-C_3N_4 heterostructure Z-scheme photo-catalyst for the elimination of Malachite Green dye from industrial water. *Optical Mater* 2025; 160: 116696.
43. Zhang LY, Cui CH, Yang SS, et al., Advanced photocatalytic degradation of highly chlorinated organic wastewater using anthraquinone-modified carbon nitride: Driven by superoxide and dichloride radicals. *Chem Eng J* 2025; 505: 159409.
44. Liu L, Tong Y, Cai X, et al., Toxic gas molecules adsorbed on the original and metal-doped two-dimensional *s*-C_3N_4: A first-principles investigation. *Colloids Surf A* 2025; 707: 135860.
45. Jourshabani M, Asrami MR, Lee B-Y, et al., Solar-active carbon nitride film integrated by free radical copolymerization for photocatalytic indoor air purification. *Small* 2025; 24: 10261.
46. Peng S, Luo X, Xu Z, et al., Design and construction of a Ag_3PO_4 NPs/protonated *g*-C_3N_4 nanosheet S-scheme heterojunction for photocatalytic degradation of C_2H_4. *Sep Purif Technol* 2025; 354: 129076.

47. Nandakumar V, Arumugam C, Karthikeyan V, et al., Graphitic carbon nitride sheets sandwiched metal oxides: A novel platform for S-scheme heterojunction generation for efficient photodegradation of volatile organics. *Colloids Surf A* 2024; 698: 134515.
48. Zhang M, Yao W, Lv Y, et al., Enhancement of mineralization ability of C_3N_4 *via* a lower valence position by a tetracyanoquinodimethane organic semiconductor. *J Mater Chem A* 2014; 2: 11432–8.
49. Zandipak R, Bahramifar N, Torabi M, et al., Synergistic effect of graphitic-like carbon nitride and sulfur-based thiazole-linked organic polymer heterostructures for boosting the photocatalytic degradation of pharmaceuticals in water. *Chem Eng J* 2024; 494: 152843.
50. Lasso-Escobar AV, Castrillon EDC, Acosta J, Modulation of electronic availability in *g*-C_3N_4 using nickel (II), manganese (II), and copper (II) to enhance the disinfection and photocatalytic properties. *Molecules* 2024; 29(16): 03775.
51. Li M, Zhang XF, Qiu J, et al., Flexible *g*-C_3N_4@bacterial cellulose aerogels for integrated indoor air purification. *Colloids Surf A* 2024; 698: 134567.
52. Chang Y, Xu KQ, Yang XL, et al., Zinc hexacyanoferrate/*g*-C_3N_4 nanocomposites with enhanced photothermal and photodynamic properties for rapid sterilization and wound healing. *Colloids Surf B* 2024; 240: 113998.
53. Chen XF, Zhang JS, Fu XZ, et al., Fe-g-C_3N_4-catalyzed oxidation of benzene to phenol using hydrogen peroxide and visible light. *J Am Chem Soc* 2009; 131: 11658–9.

8 Precursor for Diamond Synthesis and Superhard C_3N_4 Polymorph

g-C_3N_4 is a promising precursor for diamond synthesis [1]. In this chapter, the phase transformations of graphitic carbon nitride (g-CN) under high temperature and pressure conditions are explained in relation to its phase diagram, together with other high-pressure phases of carbon nitrides, including the orthorhombic and cubic phases. Since C_3N_4 polymorphs have attracted attention as superhard materials, including diamond and β-Si_3N_4, the hardnesses of the C_3N_4 polymorphs are discussed by comparing Vickers hardness and bulk modulus with diamond and β-Si_3N_4 in this chapter. The advantages of g-CN as a precursor for diamond synthesis are briefly discussed.

8.1 PHASE DIAGRAM OF g-C_3N_4 AND DIAMOND SYNTHESIS

Stability and phase transformation of g-C_3N_4 (g-CN) were studied under high temperature and pressure conditions [2,3]. Its phase diagram was determined and discussed in comparison with that of graphite, particularly graphite to diamond transformation. The phase diagram was proposed on the basis of the experiments in the pressure and temperature range of 10–25 GPa and up to 2,000°C by a multi-anvil apparatus using MgO pressure transmitting medium and Pt heater [2]. The starting g-CN powder was synthesized from 1,3,5-trichlorotriazine ($C_3N_3Cl_3$) and sodium amide ($NaNH_2$) at 220°C [4] and had an interlayer spacing d_{002} of 0.322 nm and N/C ratio of 1.37. In Figure 8.1a and b, XRD patterns are shown on the products obtained at different temperatures under 15 and 22 GPa pressures, respectively. Under 15 GPa pressure, g-CN decomposed to graphite and nitrogen gas above 1,000°C and the resultant graphite transformed to highly crystalline diamond above 1,400°C. Under 22 GPa, the transformation to diamond occurred at much lower temperature above 1,200°C, but the formed diamond had poor crystallinity, and above 1,800°C the diamond crystallinity much improved. Under lower pressure (10 GPa), only transformation to graphite occurred, and above 1,000°C, the crystallinity of graphite markedly improved, even 100 and 101 peaks being detectable. Under higher pressure (25 GPa), the transformation from graphite to diamond was accelerated, demonstrating an XRD pattern with sharp diffraction peaks. These experimental results are plotted in the pressure vs. temperature diagram to give a phase diagram for g-CN, as Figure 8.1c. The phase g-CN is stable at mild temperatures (below 600°C) up to 25 GPa pressure and decomposes to graphite and nitrogen at a temperature above 600°C–700°C under the pressure up to 15 GPa, as shown in Figure 8.1c. Above 20 GPa, g-CN transforms to diamond and nitrogen at 800°C–900°C. It must be pointed out here that the

DOI: 10.1201/9781003589051-8

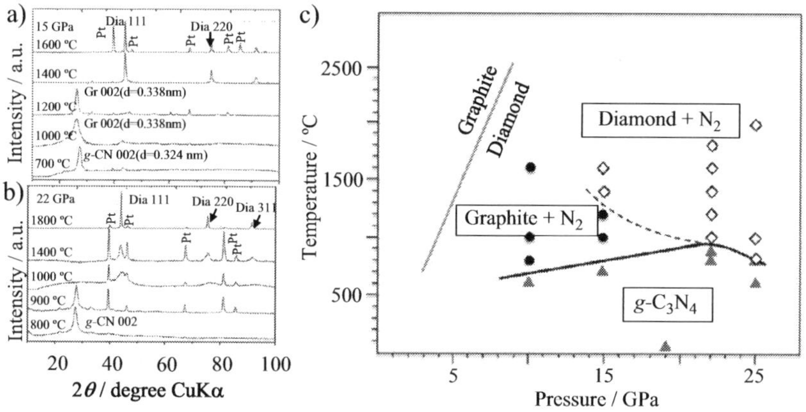

FIGURE 8.1 The products from g-C_3N_4 by the treatments at temperatures up to 2,000°C under the pressure of 10–25 GPa: XRD patterns of the products at different temperatures (a) under the pressure of 15 GPa, (b) 22 GPa, and (c) the resultant phase diagram [2].

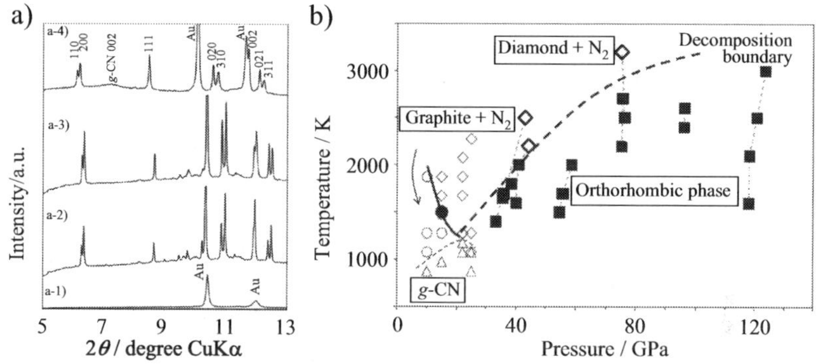

FIGURE 8.2 The products from g-C_3N_4 by the treatments at 800–3,200 K under the pressure of 5–120 GPa: (a) XRD patterns of the product obtained, (a-1) the pristine g-CN, (a-2) g-CN heated at 1,600 K before compression, (a-3) heating at 1,600 K under 30 GPa pressure, and (a-4) heated at 1,600 K and 30 GPa after de-compression, and (b) phase diagram for g-CN, where the open marks indicate the products below 30 GPa, the solid marks those above 30 GPa, the solid and broken lines indicate a kinetic boundary between graphite and diamond [3].

temperature conditions for the formation of cubic diamond via the decomposition of g-CN, shown as a broken line in Figure 8.1c, are significantly lower than those reported for direct synthesis of diamond from various carbon materials (more than 2,200°C under 15–25 GPa) [5].

The phase diagram was determined in much wider ranges in temperature and pressure by a laser-heating via a mixed Au-absorber in the pristine g-CN in a diamond anvil cell using Re foil as a gasket and either with or without the pressure-transmitting medium (NaCl), as shown in Figure 8.2b [3]. The precursor g-CN

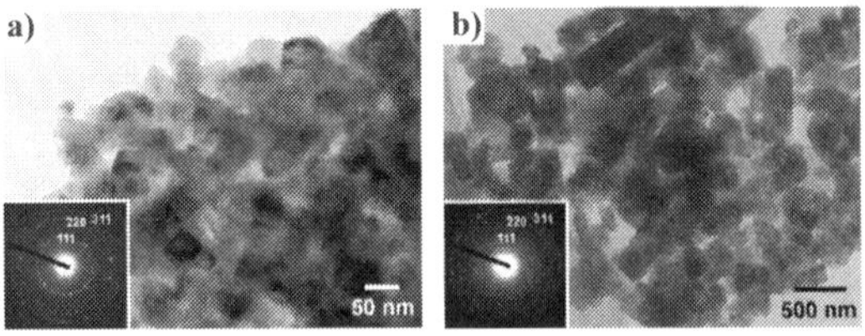

FIGURE 8.3 TEM images of the diamond (a) at 1,400°C under 15 GPa and (b) at 1,600°C under 22 GPa [2].

was synthesized through the same procedure as the above-mentioned literature [4], of which XRD showed broad and weak 002 peaks, but no *hk*0 and *hkl* peaks were detected, implying the formation of a turbostratic structure. In Figure 8.2a, the XRD patterns of the products at 1,600 K under 30 GPa pressure without mixing NaCl (a pressure-transmitting medium) are shown before and after de-compression to room temperature and ambient pressure. The product after the treatment at 1,600 K under 30 GPa (1,600 K before decompression) showed relatively sharp peaks, which were indexed by the space group $Cmc2_1$ with the orthorhombic unit cell parameter of a_0=0.76251, b_0=0.44904 and c_0=0.40424 nm, as shown in Figure 8.2a. It was noted that the unit cell parameters of this phase are practically similar to those of the CN-related phases such as $C_2N_2(NH)$ [6] and $C_2N_2(CH_2)$ [7]. This orthorhombic phase was quenchable to ambient atmosphere, as shown by the XRD pattern after de-compression. The experimental results are summarized as the phase diagram of *g*-CN, as shown in Figure 8.2b. The phase transformation behavior below 30 GPa pressure from *g*-CN to diamond was almost the same as the previous investigation (Figure 8.1c) [2]. In addition, it was found that a transformation from hexagonal *g*-CN to orthorhombic one occurs above 1,600 K and 30 GPa. The orthorhombic *g*-CN transformed directly to diamond in liquidized nitrogen formed by the decomposition of *g*-CN.

The *g*-CN has marked advantages as a precursor for diamond synthesis [2,3]. The first advantage is much milder temperature-pressure conditions for the phase transformation from graphite to diamond, as shown in Figure 8.1c and in Figure 8.2b. The second advantage is that the diamond particles were obtained as euhedral crystals and are not sintered with each other, as shown TEM images with the selected-area electron diffraction patterns in Figure 8.3a and b, probably due to its crystallization in a liquid (nitrogen) medium formed by the decomposition of the precursor *g*-CN.

8.2 C_3N_4 POLYMORPHS (SUPERHARD MATERIALS)

The *g*-CN was transformed to the orthorhombic one above 1,600 K under 30 GPa [3], of which lattice parameters were determined as a_0=0.76251, b_0=0.44904, and c_0=0.40424 nm under ambient pressure, keeping its *N*/*C* ratio with 4/3. For this orthorhombic phase formed under pressure, the unit cell volume is plotted against the pressure during compression and de-compression in Figure 8.4a. The bulk modulus

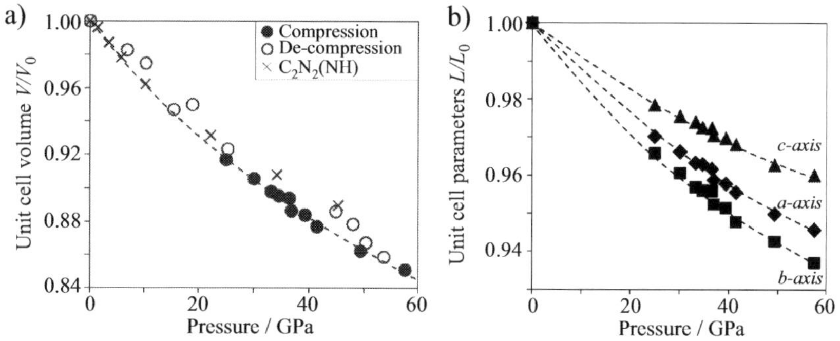

FIGURE 8.4 Compressibility of the orthorhombic C_3N_4: (a) unit cell volume during compression and de-compression processes in comparison with $C_2N_2(CH)_2$ and (b) unit cell parameters [3].

of the orthorhombic C_3N_4 was calculated to be 235 GPa, only slightly smaller than those of C_3N_4-related hydrogen-bearing carbon nitride phases (256–258 GPa), as discussed below. In Figure 8.4b, the unit cell parameters of the orthorhombic phase were plotted against the pressure applied, revealing that it is more compressible along the *a*- and *b*-axes than the *c*-axis, consistent with the case of similar carbon nitride phase $C_2N_2(CH_2)$ [8].

The orthorhombic cell with similar lattice parameters has also been reported in related carbon nitride phases such as $C_2N_2(NH)$ [6] and $C_2N_2(CH_2)$ [7]. The well-crystalized $C_2N_2(NH)$ was synthesized at a high temperature (up to 2,400°C) under a high pressure (up to 40 GPa) from a single precursor dicyandiamide (DCDA, $C_2N_4H_4$) using a CO_2-laser-heating diamond-anvil cell [6]. The phase diagram of $C_2N_2(NH)$ is shown in Figure 8.5a, which was clearly divided into two parts at a pressure of 27 GPa and a temperature of 1,600°C, Part A and B. In Part A, black amorphous products were obtained, while transparent crystals were obtained in Part B. Only a defect-wurtzite (*dwur*) structure, as shown in Figure 8.5b, was concluded on the basis of the electron diffraction EELS data, and the coordination geometry of the nitrogen sites. Its lattice parameters were calculated to be $a = 0.7536$, $b = 0.4434$, and $c = 0.4029$ nm, very similar to those of orthorhombic *g*-C_3N_4. Its bulk modulus K_0 was calculated to be 277 GPa, higher than that of β-Si_3N_4 (256 GPa), while lower than that of diamond (430–460 GPa). $C_2N_2(CH_2)$ crystal was synthesized at a temperature of 1,200–2,000 K under a pressure of 40 GPa from $C_3N_4H_xO_y$ in a diamond anvil cell with CO_2-laser-heating using Au absorber mixed in the precursor [7]. It was concluded to have the same defect-wurtzite-type structure as above-mentioned $C_2N_2(NH)$ [6]. Its lattice parameters were determined as $a = 0.7625$, $b = 0.4490$, and $c = 0.4047$ nm. By Rietveld refinement of XRD pattern of $C_2N_2(CH_2)$, two types of C were distinguished, being tetrahedrally coordinated with three N and one C, and bridging those two tetrahedra. A dense tetrahedrally bonded phase of C_2N_3H with a *dwur* structure was synthesized by laser heating of DCDA ($C_2N_4H_4$) at high pressure and temperature in a diamond anvil cell [9]. The formation process was followed via in situ synchrotron X-ray diffraction and Raman scattering. It was recoverable to ambient conditions.

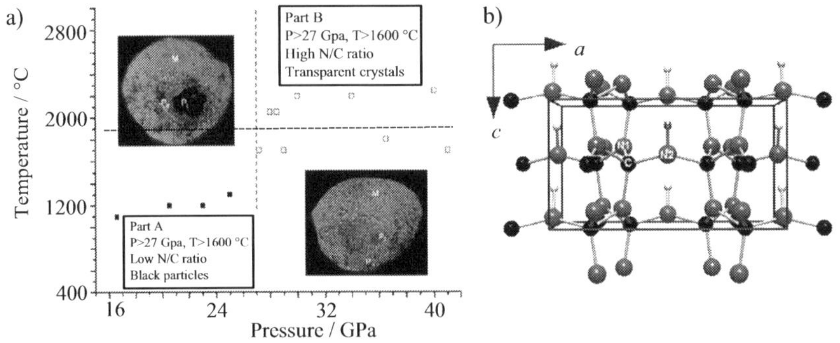

FIGURE 8.5 Orthorhombic $C_2N_2(NH)$ material: (a) phase diagram with SEM images of the particles obtained in respective parts, where P_I indicates the laser heated part and P_{II} cold rim of the particles, and M indicates the pressure-transmitting medium (NaCl), and (b) structure (a defect-wurtzite structure) [6].

The ambient pressure volume V_0 of 0.1379 nm^3 and bulk modulus K_0 of 258 GPa were in excellent agreement with density functional calculations (V_0 of 0.1347 nm^3 and K_0 of 270 GPa).

β-C_3N_4 obtained on Ni substrates by the magnetically rotated arc plasma jet CVD [10]. The arc was struck between the cone-shaped tungsten cathode and the inside wall of the graphite anode, with the magnetic field mounted externally and coaxially with the plasma torch. A nonequilibrium synthesis of β-C_3N_4 was achieved through spraying a plasma jet onto a water-cooled Ni substrate at high velocity. The flow rates of N_2 and Ar were 8 and 4 L/min, respectively. The unit cell parameters of the resultant β-C_3N_4 were calculated to be a_0 of 0.635 and c_0 of 0.248 nm. The *N/C* atomic ratios in the films grown at the substrate temperature of 600°C, 700°C, and 800°C were estimated from the Auger spectrum and increased markedly with temperature. The crystalline film of β-C_3N_4 was synthesized by a RF plasma-assisted hot filament chemical vapor deposition [11] and also by magnetron sputtering onto Si (100) substrate [12].

Although no cubic phase of C_3N_4 (c-C_3N_4) was detected in the above-mentioned investigations, the c-C_3N_4 crystalline phase with $a_0 = 0.3878$ nm was reported to be found in the coexistence of g-C_3N_4 and diamond in a diamond anvil cell with laser heating under 1,600–3,000 K and 21–38 GPa [13]. The pristine g-C_3N_4 was synthesized using the same procedure described in the above-mentioned literature [4]. The c-C_3N_4 nanoparticles were prepared in an induction thermal plasma treatment of carbon powder derived from a phenol resin [14]. The precursor carbon powder (spherical particles with 12 μm diameter) was axially injected into the center of the Ar–H_2 plasma through the carrier gas (Ar) at a feed rate of 1.4 g/min, and the powder was expected to experience over 10,000 K in the plasma. It was obtained as nm-sized particles. Relatively large ones with about 5 nm diameter were mostly in a dodecahedral shape, of which unit cell parameter was determined as 0.58 nm. A new high-pressure cubic spinel phase of C_3N_4 (cs-C_3N_4) was studied by first-principles calculations in comparison with Si_3N_4 [15].

TABLE 8.1
The Lattice Parameters *a*, *c*, Density *ρ* Bulk Modulus K_0, and Vickers Hardness H_0 of Diamond, β-Si_3N_4, and Several C_3N_4 Polymorphs

Material	Structure	Unit Cell Parameters				Density ρ (g/cm³)	Bulk modulus K_0 (GPa)	Vickers hardness H_0 (GPa)	References
		Space Group	a_0 (nm)	c_0 (nm)	Volume (nm³)				
Diamond	Cubic	$Fd\bar{3}m$	0.35363	–	0.04422	3.607	438.8	97.3	[16]
β-Si_3N_4	Hexagonal	$P63$	0.76044	0.29063	0.14555	3.44	256	14–16	[15,17]
β-C_3N_4		$P3$	0.64032	0.24053	0.08540	3.579	419.1	85.7	[16]
α-C_3N_4		$P3_{1c}$	0.64678	0.47117	0.17070	3.582	378.7	82.7	[16]
pc-C_3N_4	Pseudo-cubic	$P42m$	0.34331	–	0.04046	3.778	393.2	79.6	[16]
c-C_3N_4	Cubic	$I43d$	0.54094	–	0.15830	3.862	449.2	92.0	[16]
cs-C_3N_4	Cubic spinel	$Fd3m$	0.67138	–	0.30260	4.041	379.2	–	[16]
Ortho-C_3N_4	Orthorhombic	$Cmc2_1$	0.76251	–	–	–	235	–	[3]
C_2N_2(NH)			0.75726	0.40036	0.1409	–	277	–	[6]
C_2N_2(CH_2)			0.4490	0.4047	–	3.16	–	–	[7]
C_2N_3H			0.7618	0.4038	0.1379	–	258	–	[9]

8.3 CONCLUDING REMARKS

High-density C_3N_4 polymorphs were predicted to have a high bulk modulus and hardness comparable or exceeding that of diamond. Vickers hardness of five predicted C_3N_4 polymorphs, i.e., β-C_3N_4, α-C_3N_4, pseudocubic-C_3N_4 (*pc*-C_3N_4), cubic-C_3N_4 (*c*-C_3N_4), and cubic spinel-C_3N_4 (*cs*-C_3N_4), was calculated by using the microscopic model of hardness [16]. The results are shown in Table 8.1 by adding the data with other superhard materials, such as β-Si_3N_4 and orthorhombic C_3N_4 and $C_2N_2(NH)$. The several C_3N_4 polymorphs, including predicted ones, are superhard materials, the *c*-C_3N_4 polymorph giving the highest bulk modulus K_0, even higher than diamond, while its Vickers hardness H_0 is lower than that of diamond.

REFERENCES

1. Inagaki M, Tsumura T, Kinumoto T, et al., Graphitic carbon nitrides (*g*-C_3N_4) with comparative discussion to carbon materials. *Carbon* 2019; 141: 580–607.
2. Fang L, Ohfuji H, Shinmei T, et al., Experimental study on the stability of graphitic C_3N_4 under high pressure and high temperature. *Diam Relat Mater* 2011; 20: 819–25.
3. Kojima Y, Ohfuji H, Structure and stability of carbon nitride under high pressure and high temperature up to 125 GPa and 3000 K. *Diam Relat Mater* 2013; 39: 1–7.
4. Guo Q, Xie Y, Wang X, et al., Characterization of well-crystallized graphitic carbon nitride nanocrystallites via a benzene-thermal route at low temperatures. *Chem Phys Lett* 2003; 380: 84–7.
5. Sumiya H, Irifune T, Hardness and deformation microstructures of nanopolycrystalline diamonds synthesized from various carbons under high pressure and high temperature. *J Mater Res* 2007; 22: 2345–51.
6. Horvath-Bordon E, Riedel R, McMillan PF, et al., High-pressure synthesis of crystalline carbon nitride imide, $C_2N_2(NH)$. *Angew Chem Int Ed* 2007; 46: 1476–80.
7. Sougawa M, Sumiya T, Takarabe K, et al., Crystal structure of new carbon-nitride-related material $C_2N_2(CH_2)$. *Jpn J Appl Phys* 2011; 50: 095503.
8. Sougawa M, Sumiya T, Takarabe K, et al., Synthesis of new carbon-nitride-related materials at high pressure and temperature. *J Phys Conf Ser* 2010; 215: 012137.
9. Salamat A, Woodhead K, McMillan PF, et al., Tetrahedrally bonded dense C_2N_3H with a defective wurtzite structure: X-ray diffraction and Raman scattering results at high pressure and ambient conditions. *Phys Rev B* 2009; 80: 104106.
10. Yen T-Y, Chou C-P, Growth and characterization of carbon nitride thin films prepared by arc-plasma jet chemical vapor deposition. *Appl Phys Lett* 67, 1995; 2801–3.
11. Zhang Y, Zhou Z, Li H, Crystalline carbon nitride films formation by chemical vapor deposition. *Appl Phys Lett* 1996; 68, 634–6.
12. Wei J, Formation of g-C_3N_4 crystals at low temperature. *J Appl Phys* 2001; 89: 4099–104.
13. Ming LC, Zinin P, Meng Y, et al., A cubic phase of C_3N_4 synthesized in the diamond-anvil cell. *J Appl Phys* 2006; 99: 033520.
14. Peng Y, Ishigaki T, Horiuchi S, Cubic C_3N_4 particles prepared in an induction thermal plasma. *Appl Phys Lett* 1998; 73: 3671–3.
15. Mo S-D, Ouyang L, Ching WY, et al., Interesting physical properties of the new spinel phase of Si_3N_4 and C_3N_4. *Phys Rev Lett* 1999; 83: 5046–9.
16. He J, Guo L, Guo X, et al., Predicting hardness of dense C_3N_4 polymorphs. *Appl Phys Lett* 2006; 88: 101906.
17. Villars P. (Eds.), *β-Si_3N_4 (Si_3N_4 hex) Crystal Structure, Pauling File In: Inorganic Solid Phases, Springer Materials (Online Database)*. Springer, Heidelberg, 2024.

9 Prospectives for g-C_3N_4

Graphitic carbon nitride (g-C_3N_4) is a visible-light-active photocatalyst that can split water to produce hydrogen (H_2), a key fuel in the future *carbon-neutral society*. It is also applicable for capturing and converting CO_2, a major greenhouse gas, into valuable products such as CO and CH_4, as well as for the synthesis of H_2O_2 and NH_3. Moreover, g-C_3N_4 shows potential for degrading various environmental pollutants under ambient conditions, i.e., solar irradiation at room temperature. In addition, it is a metal-free compound and can be synthesized at relatively low temperatures, such as below 550°C, in other words, an eco-friendly process.

On the basis of these fundamental points explained in Chapters 2–8, some prospectives are described here, including the requirements for further investigations on related materials, expected further applications, etc.

9.1 PHOTOTHERMAL-ASSISTANCE FOR THE PHOTOCATALYTIC ACTIVITY

For the practical application of *g*-CN-based materials, an effective utilization of solar light seems to be required. Most of the studies on water splitting for H_2-evolution have involved dispersing the catalyst particles in water, while solar light is absorbed in water and dispersed strongly upon penetration. In the solar steam generation (SSG) process, carbon materials are widely used as solar energy absorbers and converters of solar energy into thermal energy to produce steam. It has been experimentally proven that they must be placed on the water surface [1]. Therefore, carbon foams (aerogels) are often employed as solar light absorbers, with strong attention and/or effort being paid to the steady supply of water either automatically through the carbon foam or intentionally through a hydrophilic path, such as cellulose string.

In the cases of H_2-evolution and CO_2-reduction using *g*-CN-based materials under solar light irradiation, photothermal assistance (activation) of photocatalytic activities has been pointed out [2]. From this viewpoint, the investigations into *g*-CN-coated porous carbon [3] and cellulose [4] seem to be suggestive. Both literatures suggest that the photocatalyst *g*-CN does not necessarily have to be its own particle and a coated thin layer of *g*-CN can express sufficient photocatalytic activities. In the former study, a hierarchically porous carbon was selected, which cannot be expected to float on the water surface. To fabricate a water-floating *g*-CN-coated carbon foam, where the *g*-CN coating enables photocatalytic water splitting into H_2 and O_2, and the substrate carbon foam enhances the process through photothermal effects under solar irradiation. Various carbon foams have now been developed and reported [5]. Most carbon foams can be modified hydrophilically by a simple treatment with an acid solution, such as HNO_3. Cellulose foam (aerogel) can be obtained commercially, and it is strongly hydrophilic. The *g*-CN-loading onto the cellulose foam is possible

DOI: 10.1201/9781003589051-9

simply by mixing the foam into *g*-CN water dispersion and freeze-drying [4]. The resultant *g*-CN-loaded cellulose foam can possibly be converted to *g*-CN-loaded carbon foam with reticulated texture. The resultant *g*-CN-loaded cellulose foam was applied as an air filter [4]. Recently, the composite foam fabricated by loading the composite Ag_2CrO_4/*g*-CN onto a polyurethane (PU) foam using polyvinyl alcohol (PVA) linking agent (the composite Ag_2CrO_4/*g*-CN/PVA/PU) simultaneously exhibited photothermal and photocatalytic activities [6]. Under 1.0 kW/m^2 visible light irradiation, the water evaporation rate was measured at 3.19 kg/m^2·h and a complete photocatalytic degradation of Rhodamine B (RhB) in an aqueous solution with 4.0×10^{-4} mol/L concentration was achieved within 40 min, and a maximum thermoelectric output power of 0.25 W/m^2 was obtained. By increasing the light intensity to 2.0 kW/m^2, the evaporation rate was accelerated to 8.52 kg/m^2·h, and the thermoelectric power increased to 1.1 W/m^2. On the composite foam, in addition, antibacterial activity against *E coli* was experimentally confirmed.

g-CN sponge (foam) (*s*-CN) was prepared by soaking a melamine sponge into a saturated aqueous solution of urea, freeze-drying, and then heat treating at 550°C for 4 h in N_2 [7]. Onto the resultant black sponge, Bi_2WO_6 (BWO) nanoflakes were loaded via hydrothermal treatment of *s*-CN in BWO dispersion with different BWO contents (10–50 wt%) at 180°C for 24 h to produce the BWO_X/*s*-CN composite (X = 10%–50%). The composite could be floating on the water. Under visible light irradiation, it delivered a high degradation rate of tetracycline in water, reaching 98.2% after 90 min of reaction. It was regrettable that the paper did not present the experimental results by demonstrating the benefits of the fact that the photocatalyst is floating on the water. Loading of black *g*-CN (*black*-CN) onto chitosan hydrogel (CH) with natural loofah (LF) was also performed by dispersing *black*-CN powder in a 0.2 M acetic acid solution of CH with LF (*black*-CH/LF/CS composite hydrogel) [8]. The *black*-CN was obtained by calcining a mixture of urea and tetrachlorofluorescein ($C_{20}H_8C_{14}O_5$) at 520°C for 2 h. The resultant composite hydrogel was floating on the water surface and promoted water evaporation and photodegradation of tetracycline (TC) to produce pure water under visible light irradiation. Its water evaporation rate reached 2.1 kg/m^2·h and TC-degradation efficiency was maintained at 91.4%. Chemically exfoliated *g*-CN thin flakes were refluxed in THF solution of 4-(tri-fluoromethyl)benzyl (TFMB) (80.5 g/mL) to prepare TFMB-loaded *g*-CN composite flakes [9]. The resultant composite flakes were highly amphiphilic and could be dispersed in an aqueous hydrazine solution to reduce nitrobenzene successfully to aniline via phase-boundary photocatalysis.

9.2 APPLICATION EXPANSION TO BIOMEDICAL FIELD

Here, a possibility to apply the *g*-C_3N_4-based materials to the biomedical field has been demonstrated in several studies, as mentioned below.

9.2.1 Emerging Chemicals

Various modified *g*-CNs and *g*-CN-based nanocomposites have recently been reported for use in the biomedical field, due to their abilities to degrade various antibiotics [10–15]. Additionally, *g*-CN containing N-vacancies (V_{N2C}-CN) has been

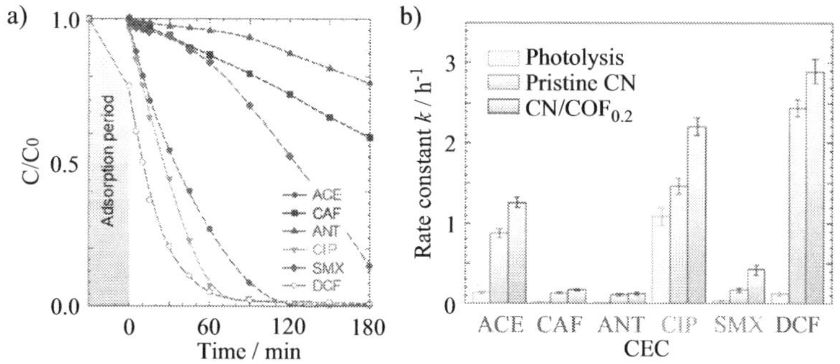

FIGURE 9.1 Photo-degradation performances of the composite $COF_{0.2}$/CN for various CECs in comparison with the pristine CN and a simple photolysis: (a) adsorption and degradation kinetics and (b) degradation rates for various CECs [19].

shown to have the capability of a rapid photodegradation removal of various emerging organic pollutants, including acetaminophen, bisphenol A, tetracycline hydrochloride, levofloxacin, and methylparaben [16–18]. An S-based COF loaded with *g*-CN sheets was prepared in a microwave-assisted process to prepare the composites (COF_X/CN) [19]. As shown in Figure 9.1a and b, the composite exhibited a high photodegradation performance for various organic contaminants of emerging concern (CECs): acetaminophen (ACE), caffeine (CAF), antipyrine (ANT), ciprofloxacin (CIP), sulfamethoxazole (SMX), and diclofenac (DCF).

9.2.2 Antibacterials

A *g*-CN-based nanocomposite with zinc hexacyanoferrate (ZHF) nanospheres (ZHF/CN) exhibited enhanced antibacterial photocatalytic activity under dual-light irradiation (420 + 808 nm) to be applied for rapid sterilization and wound healing [20]. It was also shown that the g-C_3N_4 could be coated by polydopamine (PDA) up to five times to form composites PDA/CNs, by retaining the fluorescence properties, although the fluorescence emission was reduced with increasing coating [21]. Complete PDA coating could significantly improve biomedical application potential. *g*-CN and PDA-CN can easily be protonated by a simple acid treatment, such as HCl, which also reduces bacterial growth. It was further confirmed that an improved biomedical utilization of PDA-coated material. Due to a PDA coating layer on an optically active 2D CN, for example, it may be able to be used in the treatment of neurogenerative diseases such as Parkinson's and Alzheimer's.

9.2.3 Cancer Therapy

The Ag_2S/CN nanoparticles were reported to work as a photothermal therapy (PTT) agent [22]. They exhibited a high photothermal conversion efficiency reaching 31.28% under 808 nm laser irradiation. The laser irradiation of AgS_2/CN composite nanoparticles resulted in a significant increase in reactive oxygen species generation,

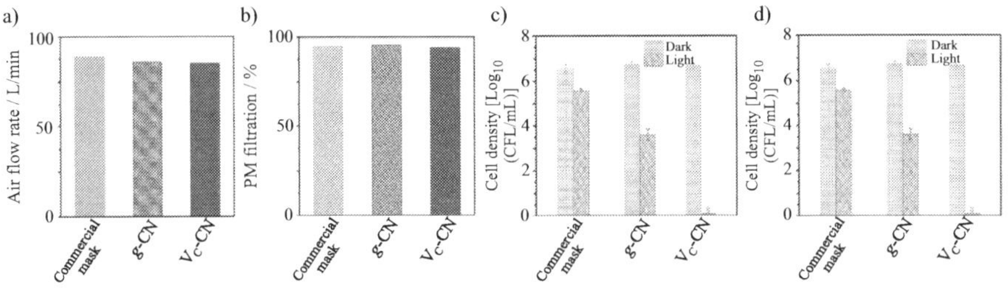

FIGURE 9.2 The V_C-CN-coated mask in comparison with the pristine commercial and *g*-CN-coated masks: (a) air flow rate, (b) particle filtration rate, bacteria survival rates for (c) *E. coli* and (d) *S. aureus* cells [23].

including· O_2^- and −OH, which could trigger the elimination of tumor cells, by confirming through cellular and animal experiments.

9.2.4 Air Filter

The flexible *g*-CN-loaded bacterial cellulose aerogels (foams) were successfully applied to an air filter for indoor air purification [4]. It exhibited high performances not only for light-driven formaldehyde degradation but also particulate matter (PM) filtration, resulting in a removal efficiency of 95.6% for formaldehyde, and a filtration efficiency of 99.3% for PM-2.5 and 99.6% for PM-10, respectively. Its aerogels also show good long-term stability, reusability, and antibacterial properties.

An antipathogen photocatalytic mask with *g*-CN containing C-vacancies (V_C-CN) nanosheets was prepared [23]. The mask was composed of V_C-CN-coating on the non-woven fabrics of the outer layer of the mask, providing effective and long-term protection against harmful pathogens when exposed to light. As shown in Figure 9.2a and b, a negligible air flow rate decrement and no obvious change in the particle rejection performance could be found, respectively, suggesting the negligible change in filtration ability even after the coating of CNs. As shown in Figure 9.2c and d, under visible light irradiation for 45 min, negligible concentrations of both bacteria, *E. coli* and *S. aureus*, were found on the mask coated with V_C-CN, while the pristine mask had almost no bacterial killing ability and the *g*-CN-coated mask exhibited certain ability.

In addition, a colloidal *g*-CN modified with hydroxyl groups was successfully applied for spray-coating on the cement-based materials [24]. On the CN-coated cement block, the photocatalytic degradation of RhB was investigated, showing a maximum RhB-degradation ratio of 75.1%.

9.3 INVESTIGATION EXPANSION TO WHOLE CARBON NITRIDE FAMILY

Graphitic carbon nitride (*g*-C_3N_4) is the most widely studied polymeric carbon nitride and is discussed in detail across eight chapters. Another stoichiometric carbon nitride, C_3N_5, is also discussed in this book in comparison with *g*-C_3N_4 [25]. These two carbon nitrides employ the repetitive units of six-membered rings (SMRs). In addition, C_4N_3 and C_3N must be mentioned as the members of SMR-based carbon nitrides. In Figure 9.3, the molecular structures of these four members are illustrated with highlighting of different bonds from C_3N_4 by broken black circles, the C_3N_5 may be called as N-rich C_3N_4-framework, and the last two (C_4N_3 and C_3N) as the C-rich C_3N_4-frameworks. C_4N_3 and C_3N, including C_3N_4, may be considered as the graphene derivatives with partial replacement of the C-atom by an N-atom and may be called the graphene family.

Carbon nitride framework with a C_3N_5 stoichiometry was synthesized by thermal deammoniation of 2,5,8-trihydrazino-s-heptazine (melem hydrazine, MH) at 450°C [26]. It has photocatalytic activity comparable to g-CN, as discussed above. C_3N_5-blended with $\left(CH_3NH_3^+\right)_x\left(CH(NH_2)_2^+\right)_{1-x}Pb(I_{0.85}Br_{0.15})_3$ (i.e., $MA_xFA_{1-x}Pb(I_{0.85}Br_{0.15})_3$) perovskite active layer in a prototype solar cell achieved an open circuit photovoltage as high as 1.3V and power conversion efficiency (PCE) as high as 16.7%, being higher than g-C_3N_4-blended. Mesoporous C_3N_5 was also prepared by KBr templating [27]. Bulk mesoporous C_3N_5 was also synthesized from amino-1H-tetrazole at 400°C. Using mesoporous silica and graphene-mesoporous silica hybrid as a template, ordered mesoporous C_3N_5 and its hybrid with graphene were prepared, respectively, both delivering enhanced oxygen reduction reaction (ORR) performance [28]. Carbon nitride framework with a C_3N stoichiometry was synthesized by polymerization of 2,3-diaminophenazine under hydrothermal conditions [29]. It consists of a 2D honeycomb lattice with a homogeneous distribution of N atoms, where both N and C atoms show a hexagonal symmetry. It is a semiconductor with an indirect bandgap of 0.39eV. FET device made of a single-layer C_3N displays an on-off current ratio reaching 5.5×10^{10}.

A layered two-dimensional network structure that possesses evenly distributed holes and nitrogen atoms, as well as a C_2N stoichiometry in its basal plane, was synthesized [30,31]. Its synthesis process is illustrated in Figure 9.4a, being a simple wet chemical reaction between hexaaminobenzene (HAB) trihydrochloride and hexaketocyclohexane (HKH) octahydrate at 175°C for 8h. The product was highly crystalline 2D flat sheets, as coded as C_2N-h2D crystals, which exhibited a shiny metallic reflection and could be easily cast into a thin film with a thickness of approximately 330nm, as shown in Figure 9.4c and d. A FET device using its film exhibited an on/off ratio of 10^7, and its bandgap was theoretically calculated and experimentally determined as approximately 1.70 and 1.96eV, respectively. C_3N_3 was prepared from cyanuric chloride on a copper surface under nonvacuum conditions. It has a good

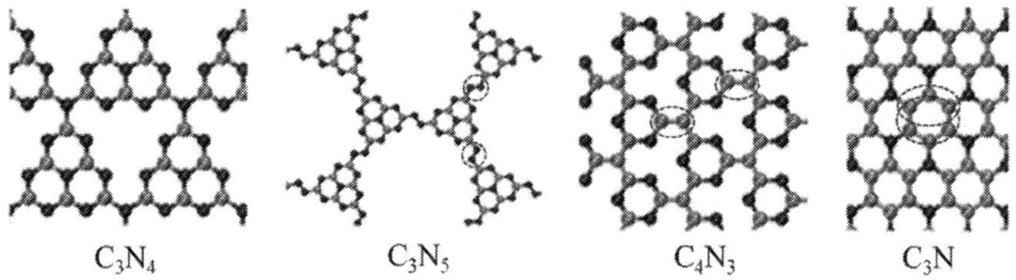

FIGURE 9.3 Carbon nitrides composed of six-membered rings [25].

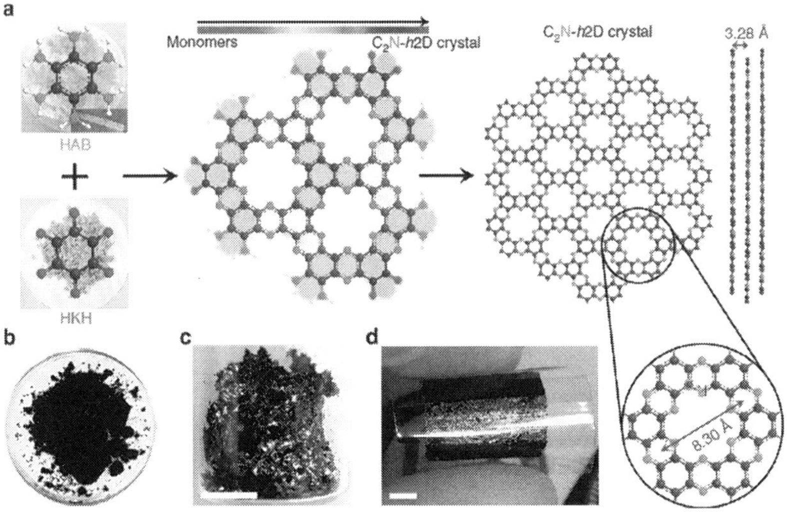

FIGURE 9.4 Layered carbon nitride with C_2N stoichiometry: (a) synthesis process, and digital photos of (b) the as-synthesized, (c) the solution cast on a SiO_2 surface, and (d) casted film transferred onto a PET substrate [30].

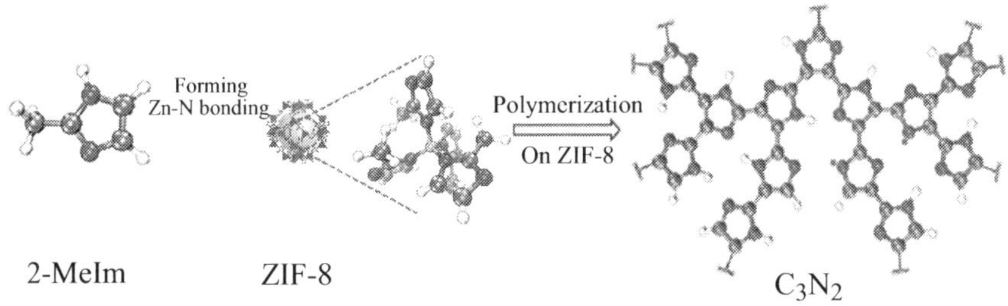

FIGURE 9.5 Synthesis and structure of C_3N_2 [25].

photoelectrochemical activity for water splitting and can be exfoliated to 2D polymeric films [32].

In addition, the framework of five-membered rings (FMRs) is also fascinating and prevalent from a fundamental point of view. One example of an FMR-based carbon nitride is C_3N_2, which was synthesized by stabilizing the precursor 2-methylimidazole (2-MeIm) with a 195°C boiling point on the surface of ZIF-8, followed by heat treatment at 600°C in N_2, as illustrated in Figure 9.5 [25].

REFERENCES

1. Toyoda M, Inagaki M, Carbon materials for solar steam-generation. *Carbon* 2023; 214: 118373.
2. Kuang L, Chen Z, Yan Y, et al., Research progress of *g*-C_3N_4-based materials for photothermal-assisted photocatalysis. *Int J Hydrogen Energy* 2024; 87: 20–49.

3. Vorokhta M, Kusdhany MIM, Švábová M, et al., Hierarchically porous carbon foams coated with carbon nitride: Insights into adsorbents for pre-combustion and post-combustion CO_2 separation. *Sep Purif Technol* 2025; 354(5): 129054.
4. Li M, Zhang XF, Qiu J, et al., Flexible g-C_3N_4@bacterial cellulose aerogels for integrated indoor air purification. *Colloids Surf A* 2024; 698: 134567.
5. Inagaki M, Toyoda M, *Carbon foams: Syntheses and Novel Applications.* Tsinghua University Press, Beijing, China (in press).
6. Wang H, Wang, T Zhu Z, et al., Construction of multifunctional photothermal/photocatalytic materials based on the principle of three primary colors: A case study of g-C_3N_4/Ag_2CrO_4. *Carbon Energy* 2025; 7(6) e711.
7. Cao D, Su N, Wang X, et al., Construction of unique floating Bi_2WO_6/g-C_3N_4 S-scheme heterojunction to promote photocatalytic activity. *J Environ Chem Eng* 2024; 12(3): 112939.
8. Gao X, Sun L, Hao P, et al., Construction of black g-C_3N_4/loofah/chitosan hydrogel as an efficient solar evaporator for desalination coupled with antibiotic degradation. *Sep Purif Tchnol* 2025; 355(A): 129615.
9. Yao C, Yuan A, Wang Z, et al., Amphiphilic two-dimensional graphitic carbon nitride nanosheets for visible-light-driven phase boundary photocatalysis. *J Mater Chem A* 2019; 7: 13071–9.
10. Liu D, Zhao C, Chen M, et al., Enhanced visible light photocatalytic performance of carbon and oxygen co-doped carbon nitride with a three-dimensional structure: Performance and mechanism study. *J Colloid Interfaces Sci* 2024; 665: 452–64.
11. Shao S, Liu X, Wang R, et al., Morphologic and microstructural modulation of graphitic carbon nitride through EDTA-2Na mediated supramolecular self-assembly route: Enhanced visible-light-driven photocatalytic activity for antibiotic degradation. *Appl Surface Sci* 2024; 669: 160501.
12. Liu Y, Chen X, Kamali M, et al., Unraveling the presence and positions of nitrogen defects in defective g-C_3N_4 for improved organic photocatalytic degradation: Insights from experiments and theoretical calculations. *Adv Funct Mater* 2024; 34: 2405741.
13. Dou L, Huang W, Huang Y, et al., Carbon-rich g-C_3N_4 nanotubes with highly dispersed cobalt as a catalyst to degrade sulfamethoxazole in the vis-PMS system: Nonradical oxidation role. *Ind Eng Chem Res* 2024; 63(13): 5697–709.
14. Song Y, Zhang Y, Wu Y, et al., Construction of a photo-assisted peroxymonosulfate activation system comprised of iron phthalocyanine/O-doped carbon nitride for efficient tetracycline removal. *J Environ Chem Eng* 2024; 12(2): 112469.
15. Zhu W, Qin L, Ou Z, et al., Protonated carbon nitride cooperated with iron porphyrin metal-organic framework for boosting photocatalytic hydrogen production and degradation of tetracycline. *Fuel* 2024; 376: 132665.
16. Yue J, Yang H, Liu C, et al., Unraveling the pyridinic nitrogen vacancy in carbon nitride for photo-self Fenton-like purification of organic contaminants. *J Colloid Interface Sci* 2024; 673: 475–85.
17. Zhang X, Liu Y, Ren M, et al., Precise carbon doping regulation of porous graphitic carbon nitride nanosheets enables elevated photocatalytic oxidation performance towards emerging organic pollutants. *Chem Eng J* 2022; 433: 134551.
18. Gan W, Chen R, Zhang L, et al., Construction of S-scheme cyano-modified g-C_3N_4/TiO_2 film with boosted charge transfer and highly hydrophilic surface for enhanced photocatalytic degradation of norfloxacin. *J Mater Sci Technol* 2025; 206: 74–87.
19. Zandipak R, Bahramifar N, Torabi M, et al., Synergistic effect of graphitic-like carbon nitride and sulfur-based thiazole-linked organic polymer heterostructures for boosting the photocatalytic degradation of pharmaceuticals in water. *Chem Eng J* 2024; 494: 152843.
20. Chang Y, Xu KQ, Yang XL, et al., Zinc hexacyanoferrate/g-C_3N_4 nanocomposites with enhanced photothermal and photodynamic properties for rapid sterilization and wound healing. *Colloids Surf B* 2024; 240: 113998.

21. Sahiner M, Demirci S, Sahiner N, Polydopamine coating of graphitic carbon nitride, *g*-C_3N_4, improves biomedical application. *Biomedicines* 2024; 12(6): 1151.
22. Wang H, You J, Shi D, et al., Multifunctional Ag_2S/*g*-C_3N_4 nanoparticles for photodynamic and photothermal cancer therapy under near-infrared excitation. *ACS Appl Nano Mater* 2024; 7: 19904–14
23. Zeng Z, Zhang Q, Ye F, et al., Bioprotective respirator assembled by defective carbon nitride for long-term light triggered health protection. *Adv Sci* 2024; 2403098
24. Yang Y, Zhang S, Yan Z, et al., Improved purification efficiency and stability of photocatalytic cement-based pavement coated with colloidal graphitic carbon nitride. *Surf Interfaces* 2024; 53: 105040.
25. Yang H, Zhou Q, Fang Z, et al., Carbon nitride of five-membered rings with low optical bandgap for photoelectrochemical biosensing. *Cell Press: Chem* 2021; 7: 2708–21.
26. Kumar P, Vahidzadeh E, Thakur UK, et al., C_3N_5: a low bandgap semiconductor containing an azo-linked carbon nitride framework for photocatalytic, photovoltaic and adsorbent applications. *J Am Chem Soc* 2019; 141: 5415–36.
27. Cai W, Li K, Zhu Z, et al., Construction of microreactor-like square-hole carbon nitride boosting photocatalytic nitrogen fixation activity. *Ind Eng Chem Res* 2024; 63(29): 12852–62.
28. Kim IY, Kim S, Jin X, et al., Ordered mesoporous C_3N_5 with a combined triazole and triazine framework and its graphene hybrids for the oxygen reduction reaction (ORR). *Angew Chem Int Ed* 2018; 57: 17135–40.
29. Yang S, Li W, Ye CW, et al., C_3N-A 2D crystalline, hole-free, tunable-narrow-bandgap semiconductor with ferromagnetic properties. *Adv Mater* 2017; 29(16): 1605625.
30. Mahmood J, Lee EK, Jung M, et al., Nitrogenated holey two-dimensional structures. *Nat Commun* 2015; 6: 6486.
31. Mahmood J, Lee EK, Jung M, et al., Two-dimensional polyaniline (C_3N) from carbonized organic single crystals in solid state. *Proc Natl Acad Sci USA* 2016; 113(27): 7414–9.
32. Feng J, Li M, Large-scale synthesis of a new polymeric carbon nitride-C_3N_3 with good photoelectrochemical performance. *Adv Funct Mater* 2020; 30(23): 2001502.

Index

L

M

N

O

P

Q

R

S

T

U

W

Printed in Dunstable, United Kingdom

1385f96e-fb65-4436-a237-ac67ab326565R01